Synthesis Lectures on Information Concepts, Retrieval, and Services

Series Editor

Gary Marchionini, School of Information and Library Science, The University of North Carolina at Chapel Hill, Chapel Hill, NC, USA

This series publishes short books on topics pertaining to information science and applications of technology to information discovery, production, distribution, and management. Potential topics include: data models, indexing theory and algorithms, classification, information architecture, information economics, privacy and identity, scholarly communication, bibliometrics and webometrics, personal information management, human information behavior, digital libraries, archives and preservation, cultural informatics, information retrieval evaluation, data fusion, relevance feedback, recommendation systems, question answering, natural language processing for retrieval, text summarization, multimedia retrieval, multilingual retrieval, and exploratory search.

Weili Guan · Xuemeng Song · Xiaojun Chang ·
Liqiang Nie

Graph Learning for Fashion Compatibility Modeling

Second Edition

 Springer

Weili Guan
Monash University
Melbourne, VIC, Australia

Xiaojun Chang
University of Technology Sydney
Sydney, NSW, Australia

Xuemeng Song
Shandong University
Qingdao, China

Liqiang Nie
Harbin Institute of Technology (Shenzhen)
Shenzhen, China

ISSN 1947-945X   ISSN 1947-9468   (electronic)
Synthesis Lectures on Information Concepts, Retrieval, and Services
ISBN 978-3-031-18819-0   ISBN 978-3-031-18817-6   (eBook)
https://doi.org/10.1007/978-3-031-18817-6

Preface

Along with economic development, numerous fashion products have sprung up in virtual and physical shops, such as bags, scarves, shoes, and skirts. Undoubtedly, they do beautify our lives. Nevertheless, they also create many troubles, especially for those who lack a sense of beauty. Moreover, people are easily overwhelmed by the abundant fashion items, thus it is difficult to find the desired fashion piece to make a compatible outfit with their wardrobes. To alleviate such a problem, the fashion compatibility modeling task, which aims to estimate the matching degree of the given set of complementary fashion items, emerged. In a sense, previous studies focused on evaluating the compatibility of an outfit with only two items, such as a top and a bottom, whereas each outfit in practice may be composed of a variable number of items. For instance, either <a top, a bottom, a pair of shoes, and a bag> or <a dress, a pair of shoes, and a hat> can compose an outfit. Accordingly, recent studies have shifted to investigating the compatibility of an outfit that involves multiple items rather than only two items. With the advances in graph learning in dealing with unstructured data, it has become a natural choice for developing the outfit compatibility modeling scheme. In fact, a few pioneer research efforts have adopted graph neural networks as the model backbone to fulfill the outfit compatibility modeling. Despite their great progress, there are still several difficult challenges that have not been well addressed as follows.

1. Multiple Correlated Modalities. How to effectively model the correlations among different modalities and fully exploit the multiple modalities of fashion items with graph learning poses a key challenge.
2. Complicated Hidden Factors. Existing methods focus on coarse-grained representation learning of composing items to derive outfit compatibility. Therefore, how to uncover the hidden factors that influence the outfit's overall compatibility and achieve fine-grained compatibility modeling with graph learning is a difficult challenge.
3. Nonunified Semantic Attributes. How to fully utilize the nonunified attribute labels of fashion items as the partial supervision for fine-grained compatibility modeling is also a considerable challenge.

4. Users' Personal Preferences. The judge of outfit compatibility is subjective. Accordingly, how to achieve personalized outfit compatibility modeling with graph learning is another challenge we are facing.

Noticing this timely opportunity, in this book, we present some state-of-the-art graph learning theories for outfit compatibility modeling, to address the aforementioned challenges. In particular, we first present a correlation-oriented graph learning method for outfit compatibility modeling, to model the correlations between visual and textual modalities of items. We next introduce the modality-oriented graph learning scheme for outfit compatibility modeling, whereby both the intramodal and intermodal compatibilities between fashion items are jointly explored. To uncover the hidden factors affecting outfit compatibility, we then devise an unsupervised disentangled graph learning method, where the fine-grained compatibility is captured. Since this unsupervised method ignores the potential of item's attribute labels in guiding fine-grained compatibility modeling, we further develop a partially supervised disentangled graph learning method. To incorporate the user's personal tastes, we propose a metapath-guided heterogeneous graph learning scheme for personalized outfit compatibility modeling. Finally, we conclude the book and identify the future research directions.

Overall, this book presents the latest advances on the topic of outfit compatibility modeling, where both the general and personalized scenarios are discussed. We expect that this book can evoke more researchers to work in this exciting and challenging area. It is suitable for students, researchers, and practitioners who are interested in the fashion compatibility modeling task.

Melbourne, Australia   Weili Guan
Qingdao, China   Xuemeng Song
Sydney, Australia   Xiaojun Chang
Shenzhen, China   Liqiang Nie

Acknowledgements

This book would not have been completed, or at least not be what it looks like now, without the support of many colleagues, especially those from the GORSE Lab at Monash University and the iLearn Center at Shandong University. It is my pleasure to take this opportunity to express my appreciation for their contributions to this time-consuming book project.

First, our sincere thanks undoubtedly go to our colleagues who contributed significantly to some chapters of this book: Dr. Chung-Hsing Yeh, Dr. Yuanfang Li, Dr. Guanliang Chen from Monash University. Mr. Tianyu Su, Mr. Haokun Wen, Mr. Fangkai Jiao, Miss. Shiting Fang, and Miss. Xiaolin Chen from Shandong University. Thanks for their participation in the technical discussion of this book and their constructive feedback and comments that significantly benefit the shaping of this book.

Second, we would like to express our heartfelt gratitude to Dr. Julie Holden from Monash University, who spared no effort to polish our earlier drafts.

Third, we are very grateful to the anonymous reviewers, who read the book very carefully and gave us many insightful and constructive suggestions. Their assistance also largely improved the quality of this book.

Fourth, we sincerely extend our thanks to Morgan & Claypool publisher, especially the editor, Dr. Gary Marchionini, and the executive editor, Ms. Diane Cerra for their valuable suggestions on this book. They also helped to make publishing the book smooth and enjoyable.

Finally, our thanks would be reserved for our beloved families for their selfless consideration, endless love, and unconditional support.

June 2022

Weili Guan
Xuemeng Song
Xiaojun Chang
Liqiang Nie

Contents

1   **Introduction**   1
    1.1   Background   1
    1.2   Challenges   2
    1.3   Our Solutions   3
    1.4   Book Structure   4
    References   4

2   **Correlation-Oriented Graph Learning for OCM**   7
    2.1   Introduction   7
    2.2   Related Work   9
    2.3   Methodology   10
        2.3.1   Problem Formulation   10
        2.3.2   Multimodal Outfit Compatibility Modeling   11
    2.4   Experiment   15
        2.4.1   Experimental Settings   15
        2.4.2   Model Comparison   16
        2.4.3   Ablation Study   18
        2.4.4   Case Study   19
    2.5   Summary   19
    References   21

3   **Modality-Oriented Graph Learning for OCM**   25
    3.1   Introduction   25
    3.2   Related Work   28
    3.3   Methodology   30
        3.3.1   Problem Formulation   30
        3.3.2   Multimodal Embedding   31
        3.3.3   Modality-Oriented Graph Learning   32
        3.3.4   Outfit Compatibility Estimation   34
    3.4   Experiment   35
        3.4.1   Experimental Settings   36

           3.4.2   Model Comparison   37
           3.4.3   Ablation Study   38
           3.4.4   Modality Comparison   40
           3.4.5   Hyperparameter Discussion   41
     3.5   Summary   44
     References   45

4   Unsupervised Disentangled Graph Learning for OCM   49
     4.1   Introduction   49
     4.2   Related Work   51
     4.3   Methodology   51
           4.3.1   Problem Formulation   52
           4.3.2   Context-Aware Outfit Representation Learning   52
           4.3.3   Hidden Complementary Factors Learning   55
           4.3.4   Outfit Compatibility Modeling   56
     4.4   Experiment   57
           4.4.1   Experimental Settings   57
           4.4.2   Model Comparison   58
           4.4.3   Ablation Study   60
           4.4.4   Case Study   61
     4.5   Summary   63
     References   64

5   Supervised Disentangled Graph Learning for OCM   67
     5.1   Introduction   67
     5.2   Related Work   69
     5.3   Methodology   69
           5.3.1   Problem Formulation   70
           5.3.2   Partially Supervised Compatibility Modeling   70
     5.4   Experiment   76
           5.4.1   Experimental Settings   76
           5.4.2   Model Comparison   78
           5.4.3   Ablation Study   79
           5.4.4   Case Study   82
     5.5   Summary   85
     References   85

6   Heterogeneous Graph Learning for Personalized OCM   89
     6.1   Introduction   89
     6.2   Related Work   91
     6.3   Methodology   93
           6.3.1   Problem Formulation   93
           6.3.2   Metapath-Guided Personalized Compatibility Modeling   94

6.4 Experiment 99
    6.4.1 Experimental Settings 100
    6.4.2 Model Comparison 102
    6.4.3 Ablation Study 102
    6.4.4 Sensitivity Analysis 104
    6.4.5 Case Study 105
6.5 Summary 106
References 106

7 Research Frontiers 109
7.1 Efficient Fashion Compatibility Modeling 110
7.2 Unbiased Fashion Compatibility Modeling 111
7.3 Try-On Enhanced Fashion Compatibility Modeling 111
References 112

About the Authors

Weili Guan received her Master's degree from National University of Singapore. After that, she joined Hewlett Packard Enterprise in Singapore as a Software Engineer and worked there for around 5 years. She is currently a Ph.D. student at the Faculty of Information Technology, Monash University (Clayton Campus), Australia. Her research interests are multimedia computing and information retrieval. She has published more than 20 papers at first-tier conferences and journals, like ACM MM, SIGIR, and IEEE TIP.

Xuemeng Song is currently an Associate Professor at Shandong University, China, and IEEE senior member. She received her B.E. degree from the University of Science and Technology of China, in 2012, and her Ph.D. degree from the National University of Singapore, in 2016. She has published more than 50 papers in the top venues (e.g., IEEE TIP, IEEE TMM, ACM SIGIR, ACM MM, and ACM TOIS) and 3 books. Her research interests include information retrieval and multimedia analysis. She is an editorial board member of Information Processing & Management. She is also the program committee member of several top conferences (e.g., ACM SIGIR and MM), and reviewer for top journals (e.g., IEEE TMM, IEEE TIP, and IEEE TKDE). She won the AI 2000 Most Influential Scholar Award Honorable Mention (in the field of Multimedia) by AMiner in 2022.

Xiaojun Chang is a Professor at Australian Artificial Intelligence Institute, Faculty of Engineering and Information Technology, University of Technology Sydney. He is the Director of The ReLER Lab. He is also an Honorary Professor in the School of Computing Technologies, RMIT University, Australia. Before joining UTS, he was an Associate Professor at School of Computing Technologies, RMIT University, Australia. After graduation, he subsequently worked as a Postdoc Research Fellow at School of Computer Science, Carnegie Mellon University, Lecturer and Senior Lecturer in the Faculty of Information Technology, Monash University, Australia. He has focused his research on exploring multiple signals (visual, acoustic, and textual) for automatic content analysis in unconstrained or surveillance videos. His team has won multiple prizes from international grand challenges which hosted competitive teams from MIT, University of Maryland, Facebook AI Research (FAIR), and Baidu VIS, and aim to advance visual understanding using deep learning. For example, he won the first place in the TrecVID 2019—Activity Extended Video (ActEV) challenge, which was held by National Institute of Standards and Technology, US.

Liqiang Nie is currently the Dean of the Department of Computer Science and Technology, Harbin Institute of Technology (Shenzhen). He received his B.Eng. and Ph.D. degrees from Xi'an Jiaotong University and National University of Singapore (NUS), respectively. His research interests lie primarily in multimedia computing and information retrieval. Dr. Nie has co-authored more than 100 papers and 4 books, and received more than 15,000 Google Scholar citations. He is an AE of IEEE TKDE, IEEE TMM, IEEE TCSVT, ACM ToMM, and Information Science. Meanwhile, he is the regular area chair of ACM MM, NeurIPS, IJCAI, and AAAI. He is a member of the ICME steering committee. He has received many awards, like ACM MM and SIGIR best paper honorable mention in 2019, SIGMM rising star in 2020, TR35 China 2020, DAMO Academy Young Fellow in 2020, and SIGIR best student paper in 2021.

Introduction

1.1 Background

In modern society, clothing plays an increasingly important role in people's social lives, as a compatible outfit can largely improve one's appearance. Nevertheless, not all people grow a keen sense of aesthetics, and hence often find it difficult to assemble compatible outfits. Therefore, it is highly desired to develop automatic fashion compatibility modeling methods to help people to evaluate the compatibility of a given outfit, i.e., a set of complementary items.

Figure 1.1 shows several outfit composition examples shared online by fashion experts. As can be seen, each outfit is composed of a variable number of items. The first outfit involves four complementary items, while the second outfit involves six pieces. Moreover, items in fashion-oriented communities or e-commerce platforms are usually associated with multiple modalities, including the visual images, textual descriptions, and semantic attributes (e.g., color and material). The rich multimodal data of the tremendous volume of outfit compositions have promoted many researchers' investigations in outfit compatibility modeling.

Existing outfit compatibility modeling methods can be classified into three groups: pairwise, listwise, and graphwise modeling. The pairwise modeling [1, 2] justifies the compatibility between two items, lacking a global view of the outfit that involves multiple items. The listwise method conceives the outfit as a list of items in a predefined order and evaluates the outfit compatibility with neural networks, such as bidirectional long short-term memory (Bi-LSTM) [3, 4]. Apparently, one limitation of this branch is that there is no explicit order among composing outfit items. Graphwise modeling organizes the outfit as an item graph and employs a graph learning technique to fulfill the fashion compatibility modeling task. Recently, graph learning has become the mainstream technique for addressing the fashion compatibility modeling task.

© The Author(s), under exclusive license to Springer Nature Switzerland AG 2022 1
W. Guan et al., *Graph Learning for Fashion Compatibility Modeling*,
Synthesis Lectures on Information Concepts, Retrieval, and Services,
https://doi.org/10.1007/978-3-031-18817-6_1

(a) Composition 1   (b) Composition 2   (c) Composition 3

Fig. 1.1 Three examples of outfit compositions shared on fashion-oriented websites

1.2 Challenges

In fact, performing the outfit compatibility modeling based on graph learning is nontrivial due to the following difficult challenges.

The first challenge is the multiple correlated modalities. As mentioned, online fashion items are associated with multiple modalities, such as images, textual descriptions, and semantic attribute labels. Intuitively, as the multiple modalities actually (e.g., visual and textual modalities) serve to characterize the same item, there should be a certain latent consistency shared by different modalities. Beyond that, each modality may express some unique aspects of the given items, and the multiple modalities hence supplement each other. Therefore, how to explicitly model the consistent and complementary correlations among different modalities and fully utilize the multiple fashion item modalities via graph learning poses the first challenge.

The second challenge is that there are multiple hidden factors affecting the outfit compatibility evaluation. Existing methods estimate the overall outfit compatibility by learning the coarse-grained representation of composing items. Actually, outfit compatibility is influenced by multiple latent factors, like color, texture, and style. In other words, to judge the compatibility of an outfit, we need to consider multiple factors. Therefore, how to explore the hidden factors and capture the latent fine-grained compatibility with graph learning to promote the model performance is a difficult challenge.

The third challenge is nonunified semantic fashion item attributes. According to our observation, the item-attribute labels usually convey rich information, characterizing the key parts of the items. We thus argue that the item-attribute labels should be considered to supervise the hidden factors learning and jointly improve the model performance and interpretability. Nevertheless, the fashion item-attribute labels are not unified or aligned, i.e., different items may have different attribute labels. Thereby, how to fully employ the nonunified item-attribute labels to supervise the graph learning-based fine-grained compatibility modeling is a considerable challenge.

The last challenge is users' personal preferences. Aesthetics are subjective. In other words, people have their preferences in making their personal ideal outfits, which may be caused by their diverse growing circumstances or educational backgrounds. For instance, given the same pink shirt, women who prefer a classic style prefer to match the shirt with a homochromatic skirt and high-heeled shoes, whereas women who prefer a sporty style like to coordinate the shirt with casual jeans and white sneakers. Accordingly, how to achieve the graph-based personalized outfit compatibility modeling is another challenge we face.

1.3   Our Solutions

To address the aforementioned research challenges, we present several state-of-the-art graph learning theories for outfit compatibility modeling.

To model the correlation among different modalities, we introduce a correlation-oriented graph learning method. It first nonlinearly projects each modality (visual image and textual description) into separable consistent and complementary spaces via multilayer perceptron and then models the consistent and complementary correlations between two modalities by parallel and orthogonal regularizations. Thereafter, we strengthen the visual and textual item representations with complementary information, and further induct both the text-oriented and vision-oriented outfit compatibility modeling with graph convolutional networks (GCNs). We ultimately employ the mutual learning strategy to reinforce the final compatibility modeling performance.

Toward comprehensive compatibility evaluation, we present a modality-oriented graph learning scheme, which simultaneously takes the visual, textual, and category modalities of fashion items as input. Notably, distinguished from existing work, we deem the fashion item category information as a specific modality, similar to the visual and textual modalities, and can be directly used for evaluating outfit compatibility. Additionally, the proposed graph learning-based scheme jointly considers the intramodal and intermodal compatibilities among fashion items during the information propagation, where the former refers to the compatibility relation between the same fashion item modalities in an outfit, while the latter denotes that between different modalities.

Beyond the aforementioned two methods that only exploit the overall coarse-grained compatibility and the item representation, we then devise an unsupervised disentangled graph learning method, which can capture the hidden factors that affect the outfit compatibility and learn the overall outfit representation rather than the item representations. In particular, this method introduces multiple parallel compatibility modeling networks, each of which corresponds to a factor-oriented context-aware outfit representation modeling. It is worth noting that in each network, we propose to adaptively learn the overall outfit representation based on GCNs equipped with the multihead attention mechanism, as different items contribute differently to the outfit compatibility evaluation.

To fully utilize the items' attribute labels (e.g., color and pattern), we develop a partially supervised disentangled graph learning method, where the irregular attribute labels of fashion items are utilized to guide the fine-grained compatibility modeling. In particular, we first devise a partially supervised attribute-level embedding learning component to disentangle the fine-grained attribute embeddings from the entire visual feature of each image. We then introduce a disentangled completeness regularizer to prevent information loss during disentanglement. Thereafter, we design a hierarchical GCN, which seamlessly integrates the attribute- and item-level compatibility modeling, to enhance the outfit compatibility modeling.

To incorporate the user's personal tastes and make our compatibility modeling more practical, we propose a personalized compatibility modeling scheme. In particular, we creatively build a heterogeneous graph to unify the three types of entities (i.e., users, items, and attributes) and their relations (i.e., user-item interactions, item-item matching relations, and item-attribute association relations). We also define the user-oriented and item-oriented metapaths and propose performing metapath-guided heterogeneous graph learning to enhance the user and item embeddings. Moreover, we introduce contrastive regularization to improve the model performance.

1.4   Book Structure

The remainder of this book consists of six chapters. Chapter 2 details the proposed correlation-oriented graph learning method, which jointly models the consistent and complementary relations between the visual and textual modalities of fashion items. In Chap. 3, we introduce the modality-oriented graph learning method for outfit compatibility modeling, whereby all the visual, textual and category modalities are considered and the intramodal and intermodal compatibilities between fashion items are simultaneously investigated. In Chap. 4, we present the unsupervised disentangled graph learning method, where the fine-grained compatibility is explored. In Chap. 5, we then develop a partially supervised disentangled graph learning to boost the model performance and interpretability. In Chap. 6, we introduce the personalized outfit compatibility modeling scheme to deal with the user's personal preference for fashion items. Ultimately, we conclude this book and identify the future research directions in Chap. 7.

References

1. Han, Xianjing, Xuemeng Song, Jianhua Yin, Yinglong Wang, and Liqiang Nie. 2019. Prototype-guided Attribute-wise Interpretable Scheme for Clothing Matching. In *Proceedings of the International ACM SIGIR Conference on Research and Development in Information Retrieval*, 785–794. ACM.
2. Liu, Jinhuan, Xuemeng Song, Liqiang Nie, Tian Gan, and Jun Ma. 2020. An End-to-End Attention-Based Neural Model for Complementary Clothing Matching. *ACM Transactions on Multimedia*

Computing, Communications and Applications 15 (4): 114:1–114:16.

3. Han, Xintong, Zuxuan Wu, Yu-Gang Jiang, and Larry S. Davis. 2017. Learning Fashion Compatibility with Bidirectional LSTMs. In *Proceedings of the ACM International Conference on Multimedia*, 1078–1086. ACM.

4. Dong, Xue, Jianlong Wu, Xuemeng Song, Hongjun Dai, and Liqiang Nie. 2020. Fashion Compatibility Modeling through a Multi-modal Try-on-guided Scheme. In *Proceedings of the International ACM SIGIR Conference on Research and Development in Information Retrieval*, 771–780. ACM.

Correlation-Oriented Graph Learning for OCM

2

2.1 Introduction

In this chapter, we study how to automatically evaluate the compatibility of a given outfit that involves a variable number of items with the graph learning technique. Existing graph learning-based methods focus on exploring the visual modality of fashion items, and seldom investigate an item's textual aspect, i.e., the textual description. In fact, textual descriptions of fashion items usually contain key features, which benefit item representation learning. Notably, although some studies have attempted to incorporate the textual modality, they simply adopt early/late fusion or consistency regularization to boost performance. Nevertheless, the correlations among multimodalities are complex and sophisticated and are not yet clearly separated and explicitly modeled.

However, outfit compatibility modeling via exploiting the multimodal correlations is nontrivial considering the following facts. (1) The visual and textual modalities characterize the same item and thus should share certain consistency. As shown in Fig. 2.1a, both visual and textual modalities deliver the item's features of "color" and "category". Additionally, the user-generated text may provide complementary information to the visual image, such as the item brand "New Ace" and material "Leather" in Fig. 2.1b, yet certain features are difficult to describe using textual sentences but easy to visualize using the image, such as the stripe position in the item in Fig. 2.1b. Consistent and complementary contents are often mixed in each modality and may be nonlinearly separable. Therefore, how to explicitly separate and model them poses one challenge. (2) How to leverage the correlation modeling results to strengthen the text and vision-oriented representation of the given item forms another challenge. And (3) outfit compatibility modeling can be derived separately from vision or text-oriented representations, which characterizes the item from different angles. We argue that these two kinds of modeling share certain common knowledge on the outfit compatibility evaluation and can reinforce each other. How to mutually enhance the two

© The Author(s), under exclusive license to Springer Nature Switzerland AG 2022
W. Guan et al., *Graph Learning for Fashion Compatibility Modeling*,
Synthesis Lectures on Information Concepts, Retrieval, and Services,
https://doi.org/10.1007/978-3-031-18817-6_2

(a)   **Dark Blue Denim Shorts**

(b)   **White Leather Stripe**
                                        New Ace Sneakers

Fig. 2.1 Illustration of the consistent and complementary correlations between the visual and textual modalities. In **a**, both the text and image reflect the color (dark blue) and category (shorts) of the item. In **b**, the text reveals the item's material (leather) and brand (New Ace) that is rarely derived visually, but fails to describe the pattern (stripe) position

kinds of modeling and thus boost the final compatibility modeling result constitutes the last challenge.

To address the aforementioned challenges, we devise a comprehensive **M**ulti**M**odal **O**utfit **C**ompatibility **M**odeling scheme, MM-OCM. As shown in Fig. 2.2, MM-OCM consists of four components: (a) multimodal feature extraction, (b) multimodal correlation modeling, (c) compatibility modeling, and (d) mutual learning. The first component extracts the textual and visual features of the given item via two separate convolution neural networks (CNNs) [1] and long short-term memory (LSTM) networks [2], respectively. The reason for introducing two separate feature extractors is to facilitate later mutual learning. The second component aims to separate and model the consistent and complementary correlations. Considering the fact that these two kinds of correlations may not be separable in the original visual and textual feature spaces, we, therefore, employ the multilayer perceptrons to nonlinearly project the image/text feature into the consistent and complementary space, where the modal-modal consistency and complementarity, respectively, can be captured. In the third component, we incorporate the disengaged complementary content in the textual (visual) modality to complement the visual (textual) feature embedding and obtain the text (vision)-oriented representation. Thereafter, we build two independent graph convolutional networks to model outfit compatibility, namely text-oriented compatibility modeling (T-OCM) and vision-oriented compatibility modeling (V-OCM). Ultimately, the fourth component targets mutually transferring knowledge from one compatibility modeling to guide the other. Once MM-OCM converges, we average the compatibility scores predicted by T-OCM and V-OCM as the final result. Extensive experiments on the real-world dataset demonstrate the superiority of our MM-OCM scheme as compared to several state-of-the-art baselines. As a byproduct, we released the codes to benefit other researchers.[1]

---

[1] https://site2750.wixsite.com/mmocm.

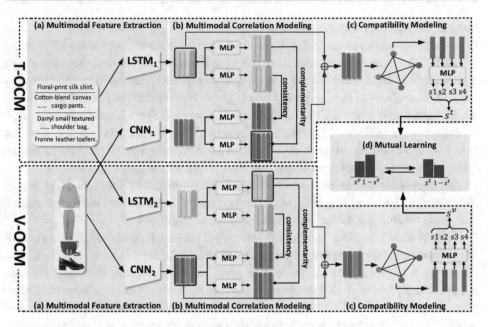

Fig. 2.2 Illustration of the proposed MM-OCM scheme. It consists of four key components: **a** multimodal feature extraction, **b** multimodal correlation modeling, **c** compatibility modeling, and **d** mutual learning

2.2 Related Work

This work is related to fashion compatibility modeling and deep mutual learning.

Fashion Compatibility Modeling. The recent flourish in the fashion industry has promoted researchers to address many fashion analysis tasks, such as clothing retrieval [3], compatibility modeling [4, 5], fashion trend prediction [6] and clothing recommendation [7, 8]. Specifically, as the key to many fashion-oriented applications, such as complementary item retrieval [9] and personal capsule wardrobe creation [10], fashion compatibility modeling has drawn great research attention. Existing fashion compatibility modeling methods can be grouped into three categories: pairwise methods [4, 11–13], listwise methods [14], and graphwise methods [15, 16]. The first category focuses on studying the compatibility between two items. For example, McAuley et al. [11] used linear transformation to map items into a latent space, where the compatibility relation between items can be measured. Following that, Song et al. [4] proposed a multimodal compatibility modeling scheme, where neural networks are used to model the compatibility between fashion items with the Bayesian personalized ranking (BPR) [17] optimization. Later, Vasileva et al. [18] studied the compatibility for outfits with multiple fashion items based on pairwise modeling, where the item category information was additionally considered. One key limitation of this category is that it lacks a global view of the outfit and rarely generates the optimal solution. The second category regards the outfit as a sequence of items in a fixed predefined order. For example, Han et al. [14] employed the Bi-LSTM network to uncover outfit compatibility. It is worth noting that the underlying assumption used by the listwise methods, i.e., the outfit can be represented as a sequence of ordered items, is questionable. Approaches in the third category model each outfit as an item graph and turn to graph neural networks [19, 20] to fulfill the outfit compatibility modeling task. For example, Cui et al. [15] proposed Node-wise graph neural networks (NGNNs) to promote item representation learning. In addition, Cucurull et al. [16] addressed the compatibility prediction problem using a graph neural network that learns to generate product embeddings conditioned on their context.

Although these studies have achieved significant success, they focus on either simply exploring the visual modality of the outfit, or considering both the visual and textual modalities while overlooking the sophisticated multimodal correlations.

Deep Mutual Learning. The idea of deep mutual learning is developed from the knowledge distillation, which was first introduced by Hinton et al. [21] for transferring the knowledge from a large cumbersome model to a small model to improve the model portability. Specifically, Hu et al. [22] designed an iterative teacher-student knowledge distillation approach, where the teacher network understands certain knowledge, while the student network iteratively mimics the teacher's solution to a certain problem to improve its performance. After that, the teacher-student knowledge distillation scheme attracted considerable attention [23, 24]. However, in many cases, it might be too difficult to obtain a teacher network with clear domain knowledge. Accordingly, Zhang et al. [25] proposed a deep mutual learning method for the classification task, where there is no explicit static teacher but an ensemble of students learning collaboratively throughout the training process. Thereafter, many researchers have investigated the deep mutual learning in various domains, such as person reidentification [26–28], domain-adapted sentiment classification [29], and deep metric learning [30]. Despite the value of mutual learning in these fields, its potential in outfit compatibility modeling has been largely unexplored, which is the major concern of this work.

2.3   Methodology

In this section, we first formulate the research problem and then detail the proposed MM-OCM scheme.

2.3.1   Problem Formulation

We deem the outfit compatibility modeling task as a binary classification problem. Assume that we have a training set Ω composed of N outfits, i.e, $\Omega = \{(O_i, y_i) \mid i = 1, \ldots, N\}$, where O_i is the ith outfit, and y_i denotes the ground truth label. We set $y_i = 1$ if the outfit

O_i is compatible, and $y_i = 0$ otherwise. Given an arbitrary outfit O, it can be represented as a set of fashion items, i.e., $O = \{o_1, o_2, \ldots, o_m\}$, where o_i is the ith item, associated with a visual image v_i and a textual description t_i. The symbol m is a variable for different outfits, considering that the number of items in an outfit is not fixed. Based on these training samples, we target learning an outfit compatibility model $\mathcal{F}$ that can judge whether the given outfit O is compatible,

$$s = \mathcal{F}\big(\{(v_i, t_i)\}_{i=1}^m | \Theta\big),\tag{2.1}$$

where Θ is a set of to-be-learned parameters of our model, and s denotes the probability the given outfit is compatible.

2.3.2 Multimodal Outfit Compatibility Modeling

Based upon the defined research problem and notations, we present the comprehensive **M**ulti**M**odal **O**utfit **C**ompatibility **M**odeling scheme, MM-OCM. As shown in Fig. 2.2, it consists of four key components: (a) multimodal feature extraction, (b) multimodal correlation modeling, (c) compatibility modeling, and (d) mutual learning.

Multimodal Feature Extraction

We first introduce the visual and textual feature extraction.

Visual Feature Extraction. To extract visual features, we utilize the CNNs, which have shown compelling success in many computer vision tasks [31–33]. To facilitate the mutual enhancement between the T-OCM and the V-OCM, which are alternatively optimized, we employ two separate CNNs to extract the visual features. Specifically, given the outfit O, the visual feature of the ith item in the outfit can be obtained as follows,

$$\begin{cases} \hat{\mathbf{v}}_i = \mathrm{CNN}_1(v_i), \\ \tilde{\mathbf{v}}_i = \mathrm{CNN}_2(v_i), \end{cases}\tag{2.2}$$

where $\hat{\mathbf{v}}_i \in \mathbb{R}^{d_v}$ and $\tilde{\mathbf{v}}_i \in \mathbb{R}^{d_v}$ refer to the visual features to be processed by the following T-OCM and V-OCM, respectively. The symbol d_v is the dimension of the extracted visual feature embedding. CNN_1 and CNN_2 denotes the corresponding CNNs for the T-OCM and V-OCM, respectively.

Textual Feature Extraction. Due to its prominent performance in textual representation learning [34, 35], we adopt LSTM to extract the textual feature of the given item.[2] Similar to the visual feature extraction, we also use two separate LSTMs, i.e., LSTM_1 and LSTM_2, to obtain the textual features for T-OCM and V-OCM, respectively. Formally, we have

$$\begin{cases} \hat{\mathbf{t}}_i = \mathrm{LSTM}_1(t_i), \\ \tilde{\mathbf{t}}_i = \mathrm{LSTM}_2(t_i), \end{cases}\tag{2.3}$$

---

[2] Before feeding into the LSTM, the text is first tokenized into standard vocabularies.

where $\hat{\mathbf{t}}_i \in \mathbb{R}^{d_t}$ and $\tilde{\mathbf{t}}_i \in \mathbb{R}^{d_t}$ refer to the text features for the following T-OCM and V-OCM, respectively. d_t is the feature dimension. To facilitate the multimodal fusion, we set $d_t = d_v = d$ in this work.

Multimodal Correlation Modeling

As illustrated in Fig. 2.1, we argue that the visual image and textual description may possess certain consistency and complementarity information. Inspired by this, instead of unreasonably fusing the general multimodal features, we propose clearly separating and explicitly modeling the consistent and complementary contents of each modality, whereby we expect the consistent content of a modality can capture the alignment information between two modalities, and the complementary one of a modality can encode the supplement information to the other modality.

In particular, we first introduce two MLPs to separate the consistent and complementary parts of each modality, respectively. Mathematically, we have

$$
\begin{cases}
\hat{\mathbf{v}}_i^s = \mathrm{MLP}_v^s\left(\hat{\mathbf{v}}_i\right), \; \hat{\mathbf{t}}_i^s = \mathrm{MLP}_t^s\left(\hat{\mathbf{t}}_i\right), \\
\hat{\mathbf{v}}_i^p = \mathrm{MLP}_v^p\left(\hat{\mathbf{v}}_i\right), \; \hat{\mathbf{t}}_i^p = \mathrm{MLP}_t^p\left(\hat{\mathbf{t}}_i\right),
\end{cases}
\tag{2.4}
$$

where $\hat{\mathbf{v}}_i^s$ and $\hat{\mathbf{v}}_i^p$ respectively denote the consistent and complementary representation of the visual modality, and $\hat{\mathbf{t}}_i^s$ and $\hat{\mathbf{t}}_i^p$ denote that of the textual modality. It is worth mentioning that the consistent and complementary parts are probably inseparable within the original low-dimensional space. After nonlinear mapping via MLPs, we can project them into a high-dimensional space, whereby the consistent and complementary parts are distinguishable.

We then argue that the consistent representations of the two modalities are parallel, and the complementary representations are orthogonal. Accordingly, to regulate the consistent and complementary representations, we use the following objective functions:

$$
\begin{cases}
\mathcal{L}_s = \sum_{i=1}^{m} \{\cos(\hat{\mathbf{v}}_i^s, \hat{\mathbf{t}}_i^s)^2 + \cos(\tilde{\mathbf{v}}_i^s, \tilde{\mathbf{t}}_i^s)^2\}, \\
\mathcal{L}_p = \sum_{i=1}^{m} \{[\cos(\hat{\mathbf{v}}_i^p, \hat{\mathbf{t}}_i^p) - 1]^2 + [\cos(\tilde{\mathbf{v}}_i^p, \tilde{\mathbf{t}}_i^p) - 1]^2\}.
\end{cases}
\tag{2.5}
$$

where $\mathcal{L}_s$ and $\mathcal{L}_p$ refer to the consistent and complementary regularizations, respectively.

Compatibility Modeling

We here first introduce the text/vision-oriented representation learning for each item, and we then present the text/vision-oriented compatibility modeling.

Text/Vision-oriented Representation Learning. Based upon the component of multimodal correlation modeling, we can derive the complementary cues of the textual (visual) modality from the visual (textual) modality. Distinguished from the consistent parts that are shared between modalities, complementarity means exclusive and supplement information.

Inspired by this, to learn comprehensive item representations and hence boost the outfit compatibility modeling performance, we introduce two multimodal fusion strategies: text-oriented multimodal fusion and vision-oriented multimodal fusion. As to the first strategy, we take the textual feature extracted by LSTM as the basis and additionally incorporate the complementary representation of the visual modality. By contrast, in the latter fusion strategy, we strengthen the visual feature extracted by CNN with the complementary representation of the textual modality. Specifically, based upon the consistent and complementary representation of each modality, we can derive the final item representations from different fusion schemes as follows,

$$
\begin{cases}
\hat{\mathbf{o}}_i = \hat{\mathbf{t}}_i + \hat{\mathbf{v}}_i^p, \\
\tilde{\mathbf{o}}_i = \tilde{\mathbf{v}}_i + \tilde{\mathbf{t}}_i^p,
\end{cases}
\tag{2.6}
$$

where $\hat{\mathbf{o}}_i$ and $\tilde{\mathbf{o}}_i$ denote the final item representation based on the text-oriented multimodal fusion and vision-oriented multimodal fusion, respectively.

Text/Vision-oriented Compatibility Modeling. Similar to previous studies, we employ a graph convolutional network (GCN) to flexibly model the compatibility of the outfit with a variable number of items. In particular, we adopt two GCNs, one for the T-OCM, and the other for the V-OCM. Regarding the limited space, we take the T-OCM as an example, since the V-OCM can be derived in the same way. In particular, for each outfit O composed of m fashion items, we first construct an indirected graph $G = (\mathcal{E}, \mathcal{R})$. $\mathcal{E} = \{o_i\}_{i=1}^{m}$ is the set of nodes, corresponding to the items of the given outfit O. Additionally, $\mathcal{R} = \{(o_i, o_j) \,|\, i, j \in [1, \ldots, m]\}$ denotes the set of edges. In this work, for each pair of items o_i and o_j in the outfit, we introduce an edge. During learning, each node o_i is associated with a hidden state vector $\mathbf{h}_i$, which keeps dynamically updated to fulfill the information propagation over the graph. For T-OCM, we initialize the hidden state vector for the ith node based on the text-oriented representation of the ith item, namely, $\mathbf{h}_i = \hat{\mathbf{o}}_i$.

The information propagation from item o_j to item o_i is defined as follows:

$$
\mathbf{m}_{j \to i} = \phi[\mathbf{W}_{pp}(\mathbf{h}_i \odot \mathbf{h}_j) + \boldsymbol{b}_{pp}],
\tag{2.7}
$$

where $\mathbf{W}_{pp} \in \mathbb{R}^{d \times d}$ and $\mathbf{b}_{pp} \in \mathbb{R}^d$ denote the weight matrix and bias vector to be learned; $\phi(\cdot)$ is a nonlinear activation function, which is set as LeakyReLU; $\mathbf{h}_i \odot \mathbf{h}_j$ accounts for the interaction between the fashion item o_i and o_j; $\odot$ is the elementwise product operation. By summarizing the information propagated from all neighbors, the hidden state vector corresponding to the item o_i can be updated as follows,

$$
\mathbf{h}_i^* = \phi(\mathbf{W}_0 \mathbf{h}_i + \mathbf{b}_0) + \sum_{o_j \in \mathcal{N}_i} \mathbf{m}_{j \to i},
\tag{2.8}
$$

where $\mathbf{W}_0 \in \mathbb{R}^{d \times d}$ and $\mathbf{b}_0 \in \mathbb{R}^d$ denote the weight matrix and bias vector to be learned; $\mathcal{N}_i$ denotes the set of neighbor nodes of node o_i and $\mathbf{h}_i^* \in \mathbb{R}^d$ is the updated hidden representation of the item o_i.

We ultimately feed the updated item representation to an MLP, consisting of two fully-connected layers, to derive its probability of being a compatible outfit as follows,

$$
\begin{cases}
s_t^i = \mathbf{W}_2 \left[\psi \left(\mathbf{W}_1 \mathbf{h}_i^* + \mathbf{b}_1 \right) \right] + \mathbf{b}_2, \\
s_t = \sigma\left(\dfrac{1}{m} \sum_{i=1}^{m} s_t^i \right),
\end{cases}
\tag{2.9}
$$

where $\mathbf{W}_1$, $\mathbf{b}_1$, $\mathbf{W}_2$, and $\mathbf{b}_2$ are the to-be-learned layer parameters. $\psi(\cdot)$ refers to the Relu active function, and $\sigma(\cdot)$ denotes the Sigmoid function to ensure the compatibility probability falling in the range of $[0, 1]$. Notably, in the same way, we can derive the compatible probability of the outfit by V-OCM, which is termed as s_v.

Mutual Learning

In a sense, regardless of the text-oriented item representation or the vision-oriented representation, i.e., $\hat{o}_i$ and $\tilde{o}_i$, both of them fuse the multimodal data of an item. Therefore, the information encoded by these two representations is largely aligned, and hence the corresponding outfit compatibility modeling yields similar outputs. Additionally, they emphasize the different aspects of the item and hence may complement each other from a global view. Therefore, the knowledge learned by one compatibility model can guide the other model. Inspired by this, we turn to the deep mutual learning knowledge distillation scheme to regularize these two compatibility modeling results, mutually reinforcing them.

Unlike the traditional teacher-student knowledge distillation network, mutual learning replaces the one-way knowledge transfer from the static pretrained teacher to the student with the mutual knowledge distillation. In particular, an ensemble of student networks is employed to learn collaboratively. In our context, the T-OCM and the V-OTM can be treated as two student networks, and optimized alternatively. Namely, in each iteration, we only train one student network, while keeping the other fixed, which temporarily acts as the teacher.

We cast the compatibility modeling as a binary classification task, and adopt the widely-used cross-entropy loss for both T-OCM and V-OCM. Accordingly, we have the objective functions,

$$
\begin{cases}
\mathcal{L}_{ce}^t = -y log(s_t) - (1 - y) log(1 - s_t), \\
\mathcal{L}_{ce}^v = -y log(s_v) - (1 - y) log(1 - s_v),
\end{cases}
\tag{2.10}
$$

where y refers to the ground truth label of the outfit O. $\mathcal{L}_{ce}^t$ and $\mathcal{L}_{ce}^v$ are the objective functions for the T-OCM and V-OCM, respectively.

To encourage the two student networks to learn from each other, we adopt the Kullback Leibler (KL) divergence loss function to penalize the distance between the evaluation results of the T-OCM and V-OCM as follows,

$$
\begin{cases}
\mathcal{L}^{v->t} = s_v log \dfrac{s_v}{s_t} + (1 - s_v) log \dfrac{(1 - s_v)}{(1 - s_t)}, \\
\mathcal{L}^{t->v} = s_t log \dfrac{s_t}{s_v} + (1 - s_t) log \dfrac{(1 - s_t)}{(1 - s_v)}.
\end{cases}
\tag{2.11}
$$

Notably, we use $\mathcal{L}_{v->t}$ for training T-OCM, and $\mathcal{L}_{t->v}$ for training V-OCM. Finally, we have

$$\begin{cases} \mathcal{L}_t = \mathcal{L}_{ce}^t + \lambda \mathcal{L}^{v->t} + \eta \mathcal{L}_s + \mu \mathcal{L}_p, \\ \mathcal{L}_v = \mathcal{L}_{ce}^v + \lambda \mathcal{L}^{t->v} + \eta \mathcal{L}_s + \mu \mathcal{L}_p, \end{cases} \quad (2.12)$$

where λ, η, and μ are tradeoff hyperparameters. $\mathcal{L}_t$ and $\mathcal{L}_v$ are the final loss functions for the T-OCM and V-OCM, respectively. Each compatibility modeling component (i.e., T-COM or V-OCM) not only learns to correctly predict the true label of the training instances, but also learns to mimic the output of the other compatibility modeling component, where the consistent and complementary regularizations are also jointly satisfied. Notably, although both $\mathcal{L}_t$ and $\mathcal{L}_v$ have consistent and complementary regularizations, i.e., $\mathcal{L}_s$ and $\mathcal{L}_p$, the parameters to be optimized are distinguished, where the regularizations in $\mathcal{L}_t$ optimize the T-OCM, while that in $\mathcal{L}_s$ aim to learn parameters of V-OCM. Once our MM-OCM is well-trained, we take the average of the predicted compatibility probabilities of the V-OCM and T-OCM as the final compatibility probability of the outfit.

2.4   Experiment

In this section, we conducted experiments over two real-world datasets by answering the following research questions.

- **RQ1**: Does MM-OCM outperform state-of-the-art baselines?
- **RQ2**: How does each module affect MM-OCM?
- **RQ3**: How is the qualitative performance of MM-OCM?

2.4.1   Experimental Settings

Datasets
To evaluate the proposed method, we adopted the Polyvore Outfits dataset [18], which is widely utilized by several fashion analysis works [36, 37]. This dataset is collected from the Polyvore fashion website. Considering whether fashion items overlap in the training, validation and testing dataset, this dataset provides two dataset versions: the nondisjoint and disjoint versions, termed Polyvore Outfits and Polyvore Outfits-D. There are a total of 68, 306 outfits in Polyvore Outfits, divided into three sets: training set (53,306 outfits), validation set (5,000 outfits), and testing set (10,000 outfits). The disjoint version, Polyvore Outfits-D, contains a total of 32,140 outfits, where 16,995 outfits are for training, 3,000 outfits are for validation, and 15,145 outfits are for testing. Each outfit in the Polyvore Outfits has at least 2 items and up to 19 items, while that in the Polyvore Outfits-D has at least 2 items and up to 16 items. The average number of items in an outfit in Polyvore Outfits and Polyvore Outfits-D is 5.3 and 5.1, respectively.

Evaluation Tasks

To evaluate the proposed model, we conducted experiments on two tasks: outfit compatibility estimation and fill-in-the-blank (FITB) fashion recommendation.

Outfit compatibility estimation: This task is to estimate a compatibility score for a given outfit. Different from the previous study [14] that generates negative outfits randomly without any restriction, we replaced each item in the positive compatible outfit with another randomly selected item in the same category, which makes the task more challenging and practical. The ratio of positive and negative samples is set to 1 : 1. The positive compatible outfits are labeled as 1, while the negative outfits are labeled as 0. Similar to previous studies[14, 37], we selected the area under the receiver operating characteristic curve (AUC) as the evaluation metric.

FITB fashion recommendation: Given an incomplete outfit and a target item annotated with the question mark, this task aims to select the most compatible fashion item from a candidate item set to fill in the blank and transform the given incomplete outfit into a compatible and complete outfit. This task is practical since people need to buy garments to match the garments they already have. Specifically, we constructed the FITB question by randomly selecting an item from a positive/compatible outfit as the target item and replacing it with a blank. We then randomly selected 3 items in the same category along with the target item to form the candidate set. The performance on this task was evaluated by the accuracy (ACC) of choosing the correct item from the candidate items.

Implementation Details

For the image encoder, we selected the ImageNet [38] pretrained ResNet18 [32] as the backbone, and modified the last layer to make the output feature dimension as 256. Regarding the text encoder, we set the word embedding size to 512, and the dimension of the hidden layer in LSTM to 256. We alternatively trained the T-OCM and V-OCM by the Adam optimizer [39] with a fixed learning rate of 0.0001, and batch size of 16. The tradeoff hyperparameters in Eq. (2.12) are set as $\lambda = \eta = \mu = 1$. In particular, we launched 10-fold cross validation for each experiment and reported the average results. All the experiments were implemented by PyTorch on a server equipped with 4 NVIDIA TITAN Xp GPUs, and the random seeds were fixed for reproducibility.

2.4.2   Model Comparison

To validate the effectiveness of our proposed scheme, we chose the following baselines for comparison.

- **Bi-LSTM** [14] takes the items in an outfit as a sequence ordered by the item category and fulfills the fashion compatibility modeling with Bi-LSTM. For a fair comparison, we only utilized the visual information.

Table 2.1 Performance comparison between our proposed MM-OCM scheme and other baselines over two datasets

Method	Polyvore Outfits		Polyvore Outfits-D	
	Compat. AUC	FITB ACC (%)	Compat. AUC	FITB ACC (%)
Bi-LSTM	0.68	42.20	0.65	40.10
Type-aware	0.87	56.60	0.78	47.30
SCE-NET	0.83	52.80	0.82	52.10
NGNN	0.75	53.02	0.68	42.49
Context-aware	0.81	55.63	0.77	50.34
HFGN	0.84	49.90	0.70	39.03
MM-OCM	**0.93**	**63.40**	**0.88**	**58.02**

- **Type-aware** [18] designs type-specific embedding spaces according to the item category, where the textual item descriptions are adopted via the visual-semantic loss.
- **SCE-NET** [36] is a pairwise method, which utilizes multiple similarity condition masks to embed the item features into different semantic subspaces. This method also considers textual information.
- **NGNN** [15] employs a GNN to address the compatibility modeling task, where the node is updated by a gate mechanism. For multimodal features, NGNN designs two graph channels, and the final compatibility score is derived as a weighted average.
- **Context-aware** [16] regards fashion compatibility modeling as an edge prediction problem, where a graph autoencoder framework is introduced. Notably, only the visual features are employed.
- **HFGN** [37] shares the same spirit with NGNN and builds a category-oriented graph, where an R-view attention map and an R-view score map are introduced to compute the compatibility score. This baseline only uses visual features.

Table 2.1 shows the performance comparison among different methods on two datasets under two tasks. From this table, we make the following observations. (1) Among all the baselines, Bi-LSTM performs the worst, which suggests that modeling the outfit as an ordered list of items is not reasonable. (2) The methods that use multimodal features gain more promising results (e.g., Type-aware on Polyvore Outfits and SCE-NET on Polyvore Outfits-D) compared to those that only utilize the visual features (i.e., HFGN and Context-aware). This implies that considering both visual and textual modalities is rewarding in the outfit compatibility modeling task. (3) MM-OCM consistently surpasses all baseline methods on the two datasets under both tasks. This indicates the advantage of our scheme that utilizes multimodal correlation modeling and mutual learning in the context of outfit compatibility modeling. Notably, we performed the ten-fold t-test between our proposed

Table 2.2 Ablation study of our proposed MM-OCM scheme on two datasets. The best results are in boldface

Method	Polyvore Outfits		Polyvore Outfits-D	
	Compat. AUC	FITB ACC (%)	Compat. AUC	FITB ACC (%)
w/o Correlation	0.91	52.91	0.87	54.47
w/o Mutual	0.92	60.80	0.86	55.62
Image_Only	0.90	57.80	0.85	52.85
Text_Only	0.79	42.28	0.74	35.45
Concat_Directly	0.91	58.67	0.79	49.73
MM-OCM	**0.93**	**63.40**	**0.88**	**58.02**

scheme and each of the baselines. We observed that all the p-values are much smaller than 0.05, and we concluded that the MM-OCM is significantly better than the baselines.

2.4.3 Ablation Study

To verify the importance of each component in our model, we also compared MM-OCM with the following derivatives.

- **w/o Correlation**: To explore the effect of multimodal correlation modeling, we removed this component by setting $\hat{\mathbf{o}}_i = \hat{\mathbf{t}}_i$ and $\tilde{\mathbf{o}}_i = \tilde{\mathbf{v}}_i$ in Eq. (2.6).
- **w/o Mutual**: To study the effect of the mutual learning component, we removed the knowledge distillation between the T-OCM and V-OCM by setting $\lambda = 0$.
- **Image_Only** and **Text_Only**: The two derivatives were set to verify the importance of visual and textual information. Specifically, for the Image_Only, we removed T-OCM by setting $\tilde{\mathbf{o}}_i = \tilde{\mathbf{v}}_i$, while for the Text_Only, V-OCM was removed by setting $\hat{\mathbf{o}}_i = \hat{\mathbf{t}}_i$.
- **Concat_Directly**: To gain more insights into our utilization of visual and textual information, we directly concatenated the visual and textual features of each item and fed them to an MLP to obtain $\hat{\mathbf{o}}_i$. Accordingly, the correlation modeling and mutual learning were simultaneously removed.

Table 2.2 shows the ablation results of our MM-OCM. From this table, we make the following observations. (1) w/o Correlation performs worse than our MM-OCM, which proves the effectiveness of the proposed multimodal consistency and complementarity modeling. (2) MM-OCM surpasses w/o Mutual, indicating that the mutual learning component is helpful for integrating the T-OCM and V-OCM by transferring knowledge between the two modules. (3) Both Image_Only and Text_Only are inferior to MM-OCM, which suggests that it is essential to consider both visual and textual information to gain better outfit compatibility modeling effects. In addition, Image_Only outperforms Text_Only remarkably, which reflects that the image contains more useful information than the text, which corresponds with the saying that "a picture is worth a thousand words". (4) compared to our MM-OCM, Concat_Directly also delivers worse performance, implying that simply fusing visual and textual features is insufficient to explore the intrinsic correlation of the two modalities. This further verifies the superiority of our strategy that models the multimodal correlation and devises two schemes of multimodal fusion. Furthermore, it can be observed that on the more challenging Polyvore Outfits-D dataset, the results of Concat_Directly are better than those of Text_Only but worse than those of Image_Only. This phenomenon indicates that an inappropriate multimodal fusion method is less effective than only utilizing the more informative modality.

2.4.4 Case Study

To gain a thorough understanding of our model, we also conducted a qualitative evaluation of our method. Figure 2.3 intuitively shows several testing examples on the outfit compatibility estimation and fill-in-the-blank tasks. From Fig. 2.3a, we observed that for the example in the first row, which contains items with consistently black color and elegant style, our MM-OCM can assign it with a high compatible probability. As for the outfit in the last row with obviously incompatible colors, e.g., green does not go well with red, our MM-OCM gives a low compatibility score. In Fig. 2.3b, we can see that our method can choose the most suitable item from the candidate set to form a compatible outfit. For the example in the first row, the outfit lacks a pair of shoes and our MM-OCM correctly selects the first item by attributing a high compatibility score. As can be seen, the selected item matches well with other items in the query. As to the example in the second row, although our method chooses the correct answer (item D), it also gives a high compatibility score to the item B, since these two items are both dark jackets of the same style. This reconfirms the compatibility modeling capabilities of our model.

2.5 Summary

In this chapter, we solved the outfit compatibility modeling problem with graph convolutional networks by exploring the multimodal correlations. In particular, we clearly separated and explicitly modeled the consistent and complementary relations between the visual and textual modalities. This was accomplished by nonlinearly projecting the consistent and complementary contents into the separable spaces, whereby they were respectively formulated by parallel and orthogonal regularizers. We then applied the complementary information to strengthen the vision- and text-oriented representations. Based upon these two kinds of representations, two compatibility modeling brunches were derived and reinforced by mutual

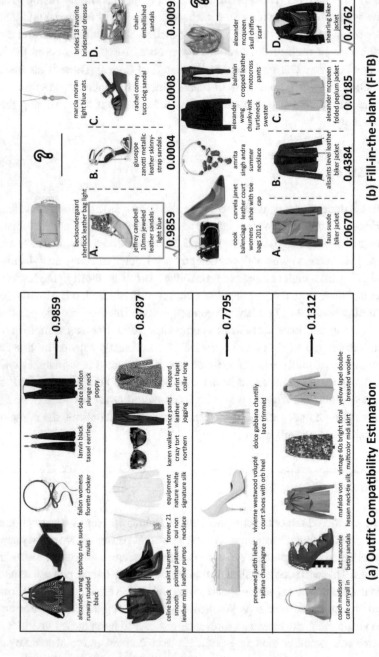

(a) Outfit Compatibility Estimation

(b) Fill-in-the-blank (FITB)

Fig. 2.3 Qualitative results of MM-OCM on **a** outfit compatibility estimation, and **b** fill-in-the-blank

learning via knowledge transfer. Extensive experiments over two benchmark datasets verified the effectiveness of our proposed MM-OCM scheme compared with several state-of-the-art baselines.

References

1. Szegedy, Christian, Vincent Vanhoucke, Sergey Ioffe, Jonathon Shlens, and Zbigniew Wojna. 2016. Rethinking the Inception Architecture for Computer Vision. In *IEEE Conference on Computer Vision and Pattern Recognition*, 2818–2826. IEEE.
2. Hochreiter, Sepp, and Jürgen. Schmidhuber. 1997. Long Short-Term Memory. *Neural Computation* 9 (8): 1735–1780.
3. Liang, Xiaodan, Liang Lin, Wei Yang, Ping Luo, Junshi Huang, and Shuicheng Yan. 2016. Clothes Co-parsing via Joint Image Segmentation and Labeling with Application to Clothing Retrieval. *IEEE Transactions on Multimedia* 18 (6): 1175–1186.
4. Song, Xuemeng, Fuli Feng, Jinhuan Liu, Zekun Li, Liqiang Nie, and Jun Ma. 2017. NeuroStylist: Neural Compatibility Modeling for Clothing Matching. In *Proceedings of the ACM International Conference on Multimedia*, 753–761. ACM.
5. Liu, Jinhuan, Xuemeng Song, Zhumin Chen, and Jun Ma. 2019. Neural Fashion Experts: I Know How to Make the Complementary Clothing Matching. *Neurocomputing* 359 (24): 249–263.
6. Gu, Xiaoling, Yongkang Wong, Pai Peng, Lidan Shou, Gang Chen, and Mohan S Kankanhalli. 2017. Understanding Fashion Trends from Street Photos via Neighbor-constrained Embedding Learning. In *Proceedings of the ACM International Conference on Multimedia*, 190–198. ACM.
7. Song, Xuemeng, Xianjing Han, Yunkai Li, Jingyuan Chen, Xin-Shun Xu, and Liqiang Nie. 2019. GP-BPR: Personalized Compatibility Modeling for Clothing Matching. In *Proceedings of the ACM International Conference on Multimedia*, 320–328. ACM.
8. Chen, Wen, Pipei Huang, Jiaming Xu, Xin Guo, Cheng Guo, Fei Sun, Chao Li, Andreas Pfadler, Huan Zhao, and Binqiang Zhao. 2019. POG: Personalized Outfit Generation for Fashion Recommendation at Alibaba iFashion. In *Proceedings of the International ACM SIGKDD Conference on Knowledge Discovery and Data Mining*, 2662–2670. ACM.
9. Wen, Haokun, Xuemeng Song, Xin Yang, Yibing Zhan, and Liqiang Nie. 2021. Comprehensive Linguistic-Visual Composition Network for Image Retrieval. In *Proceedings of the International ACM SIGIR Conference on Research and Development in Informaion Retrieval*, 1369–1378. ACM.
10. Dong, Xue, Xuemeng Song, Fuli Feng, Peiguang Jing, Xin-Shun Xu, and Liqiang Nie. 2019. Personalized Capsule Wardrobe Creation with Garment and User Modeling. In *Proceedings of the ACM International Conference on Multimedia*, 302–310. ACM.
11. McAuley, Julian J., Christopher Targett, Qinfeng Shi, and Anton van den Hengel. 2015. Image-Based Recommendations on Styles and Substitutes. In *Proceedings of the International ACM SIGIR Conference on Research and Development in Information Retrieval*, 43–52. ACM.
12. Han, Xianjing, Xuemeng Song, Jianhua Yin, Yinglong Wang, and Liqiang Nie. 2019. Prototype-guided Attribute-wise Interpretable Scheme for Clothing Matching. In *Proceedings of the international ACM SIGIR Conference on Research and Development in Information Retrieval*, 785–794. ACM.
13. Lin, Yujie, Pengjie Ren, Zhumin Chen, Zhaochun Ren, Jun Ma, and Maarten de Rijke. 2020. Explainable Outfit Recommendation with Joint Outfit Matching and Comment Generation. *IEEE Transactions on Knowledge and Data Engineering* 32 (8): 1502–1516.

14. Han, Xintong, Zuxuan Wu, Yu-Gang Jiang, and Larry S. Davis. 2017. Learning Fashion Compatibility with Bidirectional LSTMs. In *Proceedings of the ACM International Conference on Multimedia*, 1078–1086. ACM.
15. Cui, Zeyu, Zekun Li, Shu Wu, Xiaoyu Zhang, and Liang Wang. 2019. Dressing as a Whole: Outfit Compatibility Learning Based on Node-wise Graph Neural Networks. In *Porceedings of the World Wide Web Conference*, 307–317. ACM.
16. Cucurull, Guillem, Perouz Taslakian, and David Vázquez. 2019. Context-Aware Visual Compatibility Prediction. In *Proceedings of the IEEE Conference on Computer Vision and Pattern Recognition*, 12617–12626. IEEE.
17. Rendle, Steffen, Christoph Freudenthaler, Zeno Gantner, and Lars Schmidt-Thieme. 2009. BPR: Bayesian Personalized Ranking from Implicit Feedback. In *Proceedings of the International Conference on Uncertainty in Artificial Intelligence*, 452–461. AUAI Press.
18. Vasileva, Mariya I., Bryan A. Plummer, Krishna Dusad, Shreya Rajpal, Ranjitha Kumar, and David A. Forsyth. 2018. Learning Type-Aware Embeddings for Fashion Compatibility. In *European Conference on Computer Vision*, 405–421. Springer.
19. Hamilton, William L., Zhitao Ying, and Jure Leskovec. 2017. Inductive Representation Learning on Large Graphs. In *Advances in Neural Information Processing Systems*, 1024–1034. Curran Associates Inc.
20. Kipf, Thomas N., and Max Welling. 2017. Semi-Supervised Classification with Graph Convolutional Networks. In *International Conference on Learning Representations*, 1–15. OpenReview.net.
21. Hinton, Geoffrey E., Oriol Vinyals, and Jeffrey Dean. 2015. Distilling the Knowledge in a Neural Network. arXiv:1503.02531.
22. Hu, Zhiting, Xuezhe Ma, Zhengzhong Liu, Eduard H. Hovy, and Eric P. Xing. 2016. Harnessing Deep Neural Networks with Logic Rules. In *Proceedings of the Association for Computational Linguistics*, 2410–2420. ACL.
23. Zhang, Peng, Li Su, Liang Li, BingKun Bao, Pamela C. Cosman, Guorong Li, and Qingming Huang. 2019. Training Efficient Saliency Prediction Models with Knowledge Distillation. In *Proceedings of the ACM International Conference on Multimedia*, 512–520. ACM.
24. Han, Xianjing, Xuemeng Song, Yiyang Yao, Xu. Xin-Shun, and Liqiang Nie. 2020. Neural Compatibility Modeling With Probabilistic Knowledge Distillation. *IEEE Transactions on Image Processing* 29 (2020): 871–882.
25. Zhang, Ying, Tao Xiang, Timothy M. Hospedales, and Huchuan Lu. 2018. Deep Mutual Learning. In *IEEE Conference on Computer Vision and Pattern Recognition*, 4320–4328. IEEE.
26. Luo, Hao, Wei Jiang, Xuan Zhang, Xing Fan, Jingjing Qian, and Chi Zhang. 2019. AlignedReID++: Dynamically Matching Local Information for Person Re-identification. *Pattern Recognition* 94 (2019): 53–61.
27. Zhai, Yunpeng, Qixiang Ye, Shijian Lu, Mengxi Jia, Rongrong Ji, and Yonghong Tian. 2020. Multiple Expert Brainstorming for Domain Adaptive Person Re-Identification. In *Proceedings of the European Conference on Computer Vision*, 594–611. Springer.
28. Ge, Yixiao, Dapeng Chen, and Hongsheng Li. 2020. Mutual Mean-Teaching: Pseudo Label Refinery for Unsupervised Domain Adaptation on Person Re-identification. In *International Conference on Learning Representations*, 1–12. OpenReview.net.
29. Xue, Qianming, Wei Zhang, and Hongyuan Zha. 2020. Improving Domain-Adapted Sentiment Classification by Deep Adversarial Mutual Learning. In *The AAAI Conference on Artificial Intelligence*, 9362–9369. AAAI Press.
30. Park, Wonpyo, Wonjae Kim, Kihyun You, and Minsu Cho. 2020. Diversified Mutual Learning for Deep Metric Learning. In *Proceedings of the European Conference on Computer Vision*, 709–725. Springer.

31. Szegedy, Christian, Wei Liu, Yangqing Jia, Pierre Sermanet, Scott E. Reed, Dragomir Anguelov, Dumitru Erhan, Vincent Vanhoucke, and Andrew Rabinovich. 2015. Going Deeper with Convolutions. In *IEEE Conference on Computer Vision and Pattern Recognition*, 1–9. IEEE.

32. He, Kaiming, Xiangyu Zhang, Shaoqing Ren, and Jian Sun. 2016. Deep Residual Learning for Image Recognition. In *Proceedings of the IEEE Conference on Computer Vision and Pattern Recognition*, 770–778. IEEE Computer Society.

33. Liu, Meng, Liqiang Nie, Meng Wang, and Baoquan Chen. 2017. Towards Micro-video Understanding by Joint Sequential-Sparse Modeling. In *Proceedings of the ACM International Conference on Multimedia*, 970–978. ACM.

34. Chen, Yuxiao, Jianbo Yuan, Quanzeng You, and Jiebo Luo. 2018. Twitter Sentiment Analysis via Bi-sense Emoji Embedding and Attention-based LSTM. In *Proceedings of the ACM International Conference on Multimedia*, 117–125. ACM.

35. Jin, Zhiwei, Juan Cao, Han Guo, Yongdong Zhang, and Jiebo Luo. 2017. Multimodal Fusion with Recurrent Neural Networks for Rumor Detection on Microblogs. In *Proceedings of the ACM on Multimedia Conference*, 795–816. ACM.

36. Tan, Reuben, Mariya I. Vasileva, Kate Saenko, and Bryan A. Plummer. 2019. Learning Similarity Conditions Without Explicit Supervision. In *Proceedings of the IEEE International Conference on Computer Vision*, 10372–10381. IEEE.

37. Li, Xingchen, Xiang Wang, Xiangnan He, Long Chen, Jun Xiao, and Tat-Seng Chua. 2020. Hierarchical Fashion Graph Network for Personalized Outfit Recommendation. In *Proceedings of the international ACM SIGIR Conference on Research and Development in Informaion Retrieval*, 159–168. ACM.

38. Deng, Jia, Wei Dong, Richard Socher, Li-Jia Li, Kai Li, and Fei-Fei Li. 2009. ImageNet: A Large-scale Hierarchical Image Database. In *Proceedings of the IEEE Conference on Computer Vision and Pattern Recognition*, 248–255. IEEE.

39. Kingma, Diederik P., and Jimmy Ba. 2015. Adam: A Method for Stochastic Optimization. In *Proceedings of the International Conference on Learning Representations*, 1–15. OpenReview.net.

Modality-Oriented Graph Learning for OCM

3.1 Introduction

Beyond the fashion compatibility modeling, introduced in Chap. 2, which only considers the visual and textual modalities, as well as only the intramodal compatibility, this chapter presents the modality-oriented graph learning for fashion compatibility modeling, whereby both the intramodal and intermodal compatibilities between fashion items are incorporated for propagating over the entire graph. As shown in Fig. 3.1, each outfit usually consists of multiple fashion items, each of which is characterized by an image, a textual description, and category information. Therefore, to fully utilize the cues delivered by different modalities of fashion items and comprehensively model the compatibility of outfits, many research efforts [1, 2] have attempted to address the problem of outfit compatibility modeling with the multimodal information of fashion items.

Despite their remarkable performance, they suffer from the following two key limitations. (1) Prior studies focus on visual and textual modalities, and few of them utilize the fashion item category information. Additionally, these few studies [3, 4] focus on using items' categories to supervise the model learning, but fail to regard the category information as one essential input modality, i.e., comparable to the visual and textual modalities. And (2) previous efforts focus on the intramodal compatibility, i.e., the compatibility relation between the same fashion item modalities in an outfit, but overlook the intermodal compatibility, i.e., the compatibility relation between the different fashion item modalities, thereby probably causing suboptimal performance. The underlying philosophy is twofold: (a) since different modalities of an item tend to reflect the same fashion item characteristics [5, 6], incorporating the intermodal compatibility (e.g., visual-textual compatibility) can supplement the intramodal compatibility (e.g., visual-visual compatibility) and strengthen the overall compatibility estimation from an auxiliary perspective. (b) Different modalities of the same fashion item can also emphasize the different aspects of the same item. For example, the visual modality is more likely to reveal the color and pattern of the item, while the textual

© The Author(s), under exclusive license to Springer Nature Switzerland AG 2022   25
W. Guan et al., *Graph Learning for Fashion Compatibility Modeling*,
Synthesis Lectures on Information Concepts, Retrieval, and Services,
https://doi.org/10.1007/978-3-031-18817-6_3

Outfit	Image	Textual Description	Category
		sonia rykiel polka-dot satin dress	*dress*
		givenchy peplum blazer in stretch-crepe	*blazer*
		casadei red patent leather pumps	*heels*
		coach taupe leather handbag	*purse*

Fig. 3.1 Example of an outfit composition, which consists of four items. Each item has an image, textual description, and category information

modality tends to deliver its material and brand. As seen in Fig. 3.2, the given coat is visually compatible with both pairs of shoes. However, if the intermodal compatibilities between the image of the coat and the text descriptions of the two pairs of shoes are investigated, it should be easy to determine that the given top is more suitable to pair with the quilted boots rather than the canvas shoes.

Considering the outlined limitations, we propose incorporating items' category information with their content information (i.e., visual and textual modalities) and jointly model their intramodal and intermodal compatibilities to optimize the outfit compatibility

Fig. 3.2 Examples of the intermodal compatibility relation. The green and red arrows represent the compatible and incompatible relation, respectively

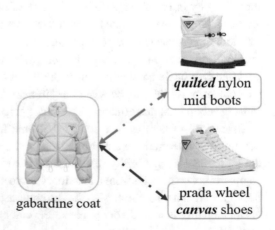

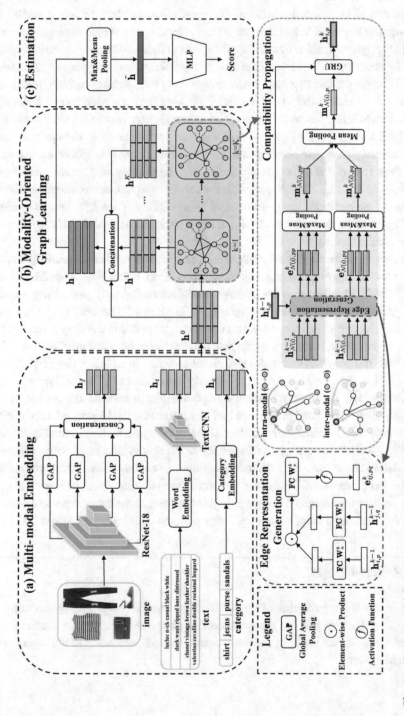

Fig. 3.3 Illustration of the proposed scheme, which consists of three modules: *multimodal embedding, modality-oriented graph learning,* and *outfit compatibility estimation*. The first module extracts the multimodal features of fashion items, and the second module refines the representation of each fashion item by absorbing its intramodal and intermodal compatibility relation with the other items. Ultimately, the third module first aggregates the composing items' representations and then uses the MLP to estimate the outfit compatibility score

modeling. However, this is nontrivial due to the following challenges. (1) Undoubtedly, the visual modality plays a pivotal role in outfit compatibility modeling. In fact, it usually delivers not only the low-level visual features (e.g., color, shape) but also the high-level visual features (e.g., style) of fashion items. Therefore, how to thoroughly explore the low-level and high-level visual features and thus benefit the compatibility modeling poses a key challenge for us. (2) Due to the fact that each outfit always comprises various fashion items, among which there is no clear order, and the matching degree between each pair of items affects the outfit compatibility, we model the outfit as an item graph. Moreover, similar to existing studies [7, 8], we resort to the GCN to fulfill the outfit compatibility modeling. Accordingly, how to effectively propagate both the intramodal and intermodal compatibilities among the fashion graph to derive the outfit compatibility also constitutes an essential challenge. (3) Essentially, one key step of outfit compatibility modeling is to learn an accurate latent representation of the outfit that can capture outfit compatibility. Therefore, how to seamlessly unify the multimodal information of fashion items to derive the latent outfit representation is another crucial challenge.

To address the aforementioned challenges, a multimodal outfit compatibility modeling scheme with modality-oriented graph learning is presented, called MOCM-MGL. As shown in Fig. 3.3, MOCM-MGL consists of three modules: *multimodal embedding*, *modality-oriented graph learning*, and *outfit compatibility estimation*. The multimodal embedding module comprises three encoders to extract the visual, textual, and category features of fashion items. In particular, multiple intermediate convolutional layers of the CNN are adopted to derive both the low-level and high-level visual features. In addition, the TextCNN [9] is utilized to embed the textual modality, and directly assign the to-be-learned embedding vector to each category. The modality-oriented graph learning module introduces a multimodal item graph for each outfit and propagates both the intramodal and intermodal compatibility relation among fashion items to refine the fashion item representations. Notably, instead of simply using the 1-D co-occurrence frequency of categories, the edge between two item nodes is defined by a multidimensional embedding to encode the complex compatibility relation between two items. Ultimately, the outfit compatibility estimation module derives the latent outfit representation by aggregating all the composing items' representations, and based on that, estimates the outfit compatibility with the multilayer perceptron (MLP) [10].

3.2   Related Work

Different from the categorization of related work on fashion compatibility modeling mentioned in Chap. 2, in this part, we group the fashion compatibility modeling studies into two categories: single-modal methods and multimodal methods, according to the input information of fashion items.

Single-modal methods only utilize the visual or textual modality of fashion items. Apparently, the visual modality plays a significant role in compatibility modeling, as many characteristics of items, such as color and shape, are encoded by visual information. Therefore, existing efforts exploit the visual information of fashion items. For example, Tangseng et al. [11] defined an outfit as a few ordered slots, corresponding to the common item categories (range from outerwear to accessory), and concatenated the visual representations of all the composing items in the outfit as the outfit representation. In addition, Cucurull et al. [12] built a graph with all fashion items in the dataset, where each node is initialized by the corresponding visual feature and receives messages from its neighborhood to learn the contextual item embedding. Apart from the visual modality, Chaidaroon et al. [13] investigated the potential of fashion item textual modality in the outfit compatibility modeling, where a text-based neural compatibility ranking model is proposed. Although great progress has been made by these works, they utilize only a single modality of fashion items and overlook the potential of combining the multimodal fashion item information.

Multimodal methods involve more than one fashion item modality. For example, Han et al. [14] proposed a bidirectional LSTM method to sequentially model the outfit compatibility by predicting the next item conditioned on previous items, where visual semantic embedding (VSE) [15] is used to capture the intermodal consistency of visual and textual modalities. This method only considers the consistency between two fashion item modalities and neglects the complementarity between them. Therefore, several researchers have been attempting to use the fusion strategy (i.e., early fusion and late fusion) to integrate the multimodal information. (1) Early fusion-based approaches typically fuse the input features extracted from each modality into a single representation before compatibility modeling [16]. For example, Tan et al. [17] fused the visual and textual features of fashion items by the elementwise product operation, while Yang et al. [18] and Sun et al. [19] directly combined the visual and textual features of each item by the concatenation operation before feeding the item feature into the compatibility modeling module. In addition, Laenen et al. [20] used the attention mechanism to fuse the visual and textual features, and projected the multimodal representations to the type-specific compatibility spaces. (2) Late fusion-based methods [7, 21] first perform the compatibility modeling directly over each modality feature, and then linearly combine the estimated outfit compatibility scores from different modalities. For example, Cui et al. [7] introduced the nodewise graph neural network (NGNN) for the outfit compatibility modeling from each modality. The overall outfit compatibility score is derived by a weighted summation of the scores obtained by the two (visual and textual) modalities. Both early fusion-based and late fusion-based methods overlook the importance of the intermodal compatibility relation between fashion items in the outfit compatibility modeling, which is the major concern of this work.

It is worth mentioning that some multimodal methods have incorporated the category information as an indicator to guide the outfit compatibility modeling. For example, Vasileva et al. [3] presented a pairwise outfit compatibility modeling scheme, where a category-specific embedding space is introduced for each pair of categories. Additionally, Wang et al. [4] learned the overall compatibility from all category-specified pairwise similarities between fashion items, and used backpropagation gradients to diagnose incompatible factors. Differently, as a major novelty, we take the category modality as one essential input modality, i.e., comparable to the visual and textual modalities, to enhance the outfit compatibility modeling performance with GCN.

3.3 Methodology

In this section, we first present the notations and problem formulation and then detail the proposed the multimodal outfit compatibility modeling scheme.

3.3.1 Problem Formulation

Since different modalities (e.g., the visual image, text description, and category) can deliver different aspects of fashion items, we propose exploring all fashion item modalities to comprehensively measure the compatibility score of outfits. Assume that we have a set of Q fashion items $\mathcal{I} = \{x_i\}_{i=1}^{Q}$, coming from N_c categories. Each fashion item $x_i \in \mathcal{I}$ is attached with a visual image, a textual description and a category, termed as f_i, t_i, and c_i, respectively. Based on these items, we can derive a set of N training outfit samples $\Omega = \{(\mathcal{O}^j, y^j) | j = 1, \ldots, N\}$, where $\mathcal{O}^j$ is the jth outfit, and y^j is the ground-truth label that indicates whether the outfit is compatible. Specifically, $y^j = 1$ denotes that the jth outfit $\mathcal{O}^j$ is compatible, and $y^j = 0$ otherwise. Each outfit can be regarded a set of fashion items, i.e., $\mathcal{O}^j = \left\{x_1^j, x_2^j, \ldots, x_{S_j}^j\right\}$, where $x_i^j \in \mathcal{I}$ denotes the ith composing item of the outfit $\mathcal{O}^j$, and S_j represents the total number of fashion items in the outfit $\mathcal{O}^j$. Notably, since each outfit can be composed of a various number of fashion items, S_j is variable. Based on these data, we aim to devise a comprehensive multimodal outfit compatibility modeling scheme $\mathcal{F}$, which can integrate the multimodal information of its composing fashion items toward the accurate outfit compatibility estimation. Mathematically, we have:

$$\hat{y}^j = \mathcal{F}(\{f_i^j, t_i^j, c_i^j\}_{i=1}^{S_j} | \Theta), \tag{3.1}$$

where f_i^j, t_i^j, and c_i^j represent the visual image, textual description, and category of the ith item of the jth outfit, respectively. Θ is a set of to-be-learned parameters, and $\hat{y}^j$ denotes the estimated compatibility score of the outfit $\mathcal{O}^j$. For brevity, we omit the superscript j of the jth outfit $\mathcal{O}^j$ in the rest of the paper.

3.3.2 Multimodal Embedding

First, we resort to the following encoders to learn the visual, textual and category representation of each fashion item.

Image Encoder. Regarding the visual image of each item, we utilize the CNN to extract its visual features. It is well known that the CNN comprises multiple convolutional layers, where the shallow layers can capture the low-level visual features, such as the color of the item, while the deep layers can capture the high-level features, such as the style of the item [22]. Since both the low-level and high-level visual features affect the compatibility among fashion items, similar to the work [4], we consider both the shallow and deep layers' outputs to learn the visual representation for each item instead of only using the final layer's output. In particular, we resort to the global average pooling operation (GAP), which has shown remarkable performance in the discriminative visual property extraction [23], to summarize the learned visual representations. Formally, given the image f_i of the fashion item x_i, we can obtain its visual feature as follows,

$$\mathbf{f}_i = \left[\text{GAP}\left(Conv^1\left(f_i\right)\right), \ldots, \text{GAP}\left(Conv^L\left(f_i\right)\right) \right], \tag{3.2}$$

where $\mathbf{f}_i \in \mathbb{R}^{d_f}$ is the visual feature of the item x_i, d_f is the dimension of the visual feature, and [,] denotes the concatenation operation. In addition, $Conv^l$ represents the lth convolutional layer used for visual encoding of CNN, and L is the total number of convolutional layers.

Text Encoder. To embed the textual description of each fashion item, we adopt the TextCNN, which has achieved astonishing success in various natural language processing tasks [24, 25]. In particular, we first represent the textual description (i.e., a sequence of words) as a matrix, each column of which refers to a word embedding learned by the pretrained word2vec. [26]. We then employ the CNN architecture to extract the semantic information of the text description of the given fashion item. Specifically, given the textual description t_i of the fashion item x_i, we obtain its textual feature as follows,

$$\mathbf{t}_i = TextCNN\left(t_i\right), \tag{3.3}$$

where $\mathbf{t}_i \in \mathbb{R}^{d_t}$ denotes the extracted textual feature, and d_t is its dimension.

Category Encoder. In addition to the visual and textual information, the category information of the composing items also plays an important role in outfit compatibility estimation. Different from previous studies that only incorporate the category information to guide the outfit compatibility modeling, we propose regarding the category as an unique input modality. To represent the discrete categories, we introduce a category embedding matrix $\mathbf{C} \in \mathbb{R}^{N_c \times d_c} = \{\mathbf{c}_1, \mathbf{c}_2, \ldots, \mathbf{c}_{N_c}\}$, where N_c is the total number of categories in the dataset, d_c is the category feature dimension, and $\mathbf{c}_k$ denotes the embedding for the kth category. Therefore, for each fashion item x_i, its category feature $\mathbf{c}_i$ can be obtained according to its category information c_i.

Ultimately, based on the above three encoders, for each fashion item x_i, we can obtain its visual feature $\mathbf{f}_i$, textual feature $\mathbf{t}_i$, and category feature $\mathbf{c}_i$.

3.3.3 Modality-Oriented Graph Learning

Since each outfit comprises a set of fashion items with no clear order, we treat each outfit as an item graph and hence resort to the GCNs to explore its outfit compatibility. In particular, for each outfit $\mathcal{O}$, we construct an item graph $\mathcal{G} = (\mathcal{V}, \mathcal{E})$, where $\mathcal{V}$ denotes the set of nodes, each of which represents a composing item, and $\mathcal{E}$ denotes the set of edges representing the compatibility relation among items. We assume that the compatibility between each pair of items should be considered in the outfit compatibility modeling, and thus make the fashion graph a complete graph. Namely, there is an edge between each pair of nodes (items).

Node Initialization
Different from the conventional methods that assign each node with a single hidden state vector, we attribute each node with three modality-oriented hidden state vectors, corresponding the three modalities. Specifically, for each fashion item x_i, we employ linear transformations to map its multimodal features into a common space to derive the modality-oriented hidden state vectors as follows,

$$\begin{cases} \mathbf{h}^0_{i,1} = \mathbf{W}_f \mathbf{f}_i + \mathbf{b}_f, \\ \mathbf{h}^0_{i,2} = \mathbf{W}_t \mathbf{t}_i + \mathbf{b}_t, \\ \mathbf{h}^0_{i,3} = \mathbf{W}_c \mathbf{c}_i + \mathbf{b}_c, \end{cases} \tag{3.4}$$

where $\mathbf{h}^0_{i,1} \in \mathbb{R}^d$, $\mathbf{h}^0_{i,2} \in \mathbb{R}^d$ and $\mathbf{h}^0_{i,3} \in \mathbb{R}^d$ are the initial hidden representations of the ith item's visual, textual and category modalities, respectively. For ease of the following presentation, without losing generality, we arrange the visual, textual and category modalities as the first, second, and third modalities of fashion items, respectively. $\mathbf{W}_f \in \mathbb{R}^{d_f \times d}$, $\mathbf{W}_t \in \mathbb{R}^{d_t \times d}$, and $\mathbf{W}_c \in \mathbb{R}^{d_c \times d}$ are the linear mapping matrices, $\mathbf{b}_f \in \mathbb{R}^d$, $\mathbf{b}_t \in \mathbb{R}^d$ and $\mathbf{b}_c \in \mathbb{R}^d$ are the biases, where d is the dimension of the hidden state representation.

Edge Representation Generation
Previous GCN-based studies [7, 8] on the outfit compatibility modeling utilize edges to indicate the graph topological information and assign each edge with a scalar. Beyond that, we model the edge between two items with a multidimensional feature rather than a one-dimensional weight, which can encode the complex compatibility relation between items. As mentioned above, in addition to the intramodal compatibility, the interaction of different modalities between fashion items can deliver the compatibility relation between fashion items. Therefore, we introduce the fine-grained edge representation $\mathbf{e}_{ij,pq}$ to capture the compatibility between the pth modality of node v_i and the qth modality of node v_j, where $p, q = 1, 2, 3$. It is worth noting that (1) when $p = q$, $\mathbf{e}_{ij,pq}$ represents the intramodal compatibility relation, and (2) when $p \neq q$, $\mathbf{e}_{ij,pq}$ represents the intermodal compatibility relation. Regarding the fine-grained edge representation generation, it is worth noting that

the order of items in each item pair does not influence the underlying compatibility relation Accordingly, in this work, we employ the symmetric elementwise product function to generate the edge representation. Specifically, we produce the fine-grained edge representation $\mathbf{e}_{ij,pq}$ for the kth propagation step as follows,

$$\mathbf{e}_{ij,pq}^{k} = \alpha \left(\mathbf{W}_{e}^{k} \left(\mathbf{W}_{n}^{k} \mathbf{h}_{i,p}^{k-1} \odot \mathbf{W}_{n}^{k} \mathbf{h}_{j,q}^{k-1} \right) + \mathbf{b}_{e}^{k} \right), \tag{3.5}$$

where $\mathbf{h}_{i,p}^{k-1}$ is the hidden representation of the pth modality of the node v_i, and $\mathbf{h}_{j,q}^{k-1}$ is the hidden representation of the qth modality of the node v_j. $\mathbf{W}_n^k \in \mathbb{R}^{d \times d}$, $\mathbf{W}_e^k \in \mathbb{R}^{d \times d_e}$, and $\mathbf{b}_e^k \in \mathbb{R}^{d_e}$ are the parameters for the edge representation generation in the kth propagation step. $\mathbf{W}_n^k$ is the weight matrix of the linear transformation to project the node embedding to latent compatibility space, while $\mathbf{W}_e^k$ is the weight matrix of the linear transformation to further compress the latent compatibility relation into a lower-dimensional space, where the compatibility relation with all the other nodes is aggregated. In particular, to facilitate the following mean pooling and max pooling-based compatibility aggregation, $d_e = d/2$. $\alpha \left(\cdot \right)$ is the ReLU activation function, and $\odot$ denotes the elementwise product operation.

Intramodal and Intermodal Compatibility Propagation
During the intramodal and intermodal compatibility propagation, we make each modality of each node absorb the fine-grained compatibility information from its connected edges to update its hidden state vector. Without losing generality, as an example, we present the compatibility aggregation process toward the pth modality of the node v_i as follows,

$$\begin{cases} \mathbf{m}_{\mathcal{N}(i),pq}^{k} = \text{AGG} \left(\left\{ \mathbf{e}_{ij,pq}^{k}, \forall j \in \mathcal{N}(i) \right\} \right), \\ \mathbf{m}_{\mathcal{N}(i),p}^{k} = \frac{1}{M} \sum_{q=1}^{M} \mathbf{m}_{\mathcal{N}(i),pq}^{k}, \end{cases} \tag{3.6}$$

where $\mathcal{N}(i)$ is the neighbors of node v_i, i.e., the nodes connected to the node v_i in the graph. $\mathbf{m}_{\mathcal{N}(i),pq}^{k} \in \mathbb{R}^d$ denotes the aggregated compatibility information from the qth modality of the node's neighbors toward the pth modality of the node v_i in the kth propagation step, while $\mathbf{m}_{\mathcal{N}(i),p}^{k} \in \mathbb{R}^d$ represents the aggregated compatibility information from all the modalities of the node's neighbors toward the pth modality of the node v_i in the kth propagation step. M is the total number of modalities, which is 3 in our context. AGG $(\cdot)$ is the aggregation function, which is implemented with both the mean and max pooling operations. Specifically, we have $\mathbf{m}_{\mathcal{N}(i),pq}^{k} =$

$$\left[\gamma_{mean} \left(\left\{ \mathbf{e}_{ij,pq}^{k}, \forall j \in \mathcal{N}(i) \right\} \right), \gamma_{max} \left(\left\{ \mathbf{e}_{ij,pq}^{k}, \forall j \in \mathcal{N}(i) \right\} \right) \right], \tag{3.7}$$

where $\gamma_{mean} (\cdot)$ and $\gamma_{max} (\cdot)$ are the mean and max pooling operations, respectively. The mean and max pooling operations are used for extracting the average and most prominent information from the connected edges, respectively.

Then, we adopt the gated recurrent unit (GRU) [27] to selectively absorb the compatibility information from the node's neighbors and the original hidden information of the node. Specifically, we define the modality representation update function for each node as follows,

$$
\begin{cases}
\mathbf{z}_{i,p}^{k} = \sigma \left(\mathbf{W}_{z}^{k} \left[\mathbf{m}_{\mathcal{N}(i),p}^{k}, \mathbf{h}_{i,p}^{k-1} \right] + \mathbf{b}_{z}^{k} \right), \\
\mathbf{r}_{i,p}^{k} = \sigma \left(\mathbf{W}_{r}^{k} \left[\mathbf{m}_{\mathcal{N}(i),p}^{k}, \mathbf{h}_{i,p}^{k-1} \right] + \mathbf{b}_{r}^{k} \right), \\
\tilde{\mathbf{h}}_{i,p}^{k} = \tanh \left(\mathbf{W}_{h}^{k} \left[\mathbf{m}_{\mathcal{N}(i),p}^{k}, \mathbf{r}_{i,p}^{k} \odot \mathbf{h}_{i,p}^{k-1} \right] + \mathbf{b}_{h}^{k} \right), \\
\mathbf{h}_{i,p}^{k} = \left(1 - \mathbf{z}_{i,p}^{k} \right) \odot \mathbf{h}_{i,p}^{k-1} + \mathbf{z}_{i,p}^{k} \odot \tilde{\mathbf{h}}_{i,p}^{k},
\end{cases}
\tag{3.8}
$$

where $\mathbf{W}_{z}^{k} \in \mathbb{R}^{2d \times d}$, $\mathbf{W}_{r}^{k} \in \mathbb{R}^{2d \times d}$, and $\mathbf{W}_{h}^{k} \in \mathbb{R}^{2d \times d}$ are weight matrices of the update function, while $\mathbf{b}_{z}^{k} \in \mathbb{R}^{d}$, $\mathbf{b}_{r}^{k} \in \mathbb{R}^{d}$ and $\mathbf{b}_{h}^{k} \in \mathbb{R}^{d}$ are biases. $\mathbf{z}_{i,p}^{k}$ and $\mathbf{r}_{i,p}^{k}$ are update gate vector and reset gate vector, respectively. $\sigma (\cdot)$ is the sigmoid activation function, and $tanh (\cdot)$ is the tanh activation function. $\mathbf{h}_{i,p}^{k}$ denotes the hidden representation of the pth modality of item x_i in the kth propagation step. As can be seen, the node update function (i.e., GRU) takes both the hidden modality representation of each node v_i and the aggregated compatibility information $\mathbf{m}_{\mathcal{N}(i),p}^{k}$ as the input. In this manner, the updated representation of the node v_i comprises not only the item intrinsic characteristics but also the compatibility relation with connected items.

3.3.4   Outfit Compatibility Estimation

After K propagation steps, we obtain a series of multimodal hidden representations of fashion item x_i, namely $\{\mathbf{h}_i^0, \ldots, \mathbf{h}_i^K\}$, where $\mathbf{h}_i^k = \left[\mathbf{h}_{i,1}^k, \mathbf{h}_{i,2}^k, \mathbf{h}_{i,3}^k \right]$, and $k = 1, \ldots, K$. Since the representations obtained at different propagation layers absorb the neighbor compatibility information at different levels, toward the comprehensive representation, we concatenate them to constitute the final representation of each fashion item x_i as follows,

$$
\mathbf{h}_i^* = \left[\mathbf{h}_i^0, \ldots, \mathbf{h}_i^K \right], i = 1, \ldots, S_j.
\tag{3.9}
$$

Thereafter, we define the final representation of the outfit based on these composing items' representations. Notably, instead of using the concatenation of all composing items' representations, we further apply a pooling layer that includes both max pooling and mean pooling operations to derive the whole outfit representation. We expect that the max pooling and mean pooling operations can capture the most prominent and the overall features of all composing items' hidden states, respectively. Specifically, we obtain the final embedding for each outfit $\mathcal{O}$ as follows,

$$
\tilde{\mathbf{h}} = \left[\gamma_{mean} \left(\{\mathbf{h}_i^*, \forall v_i \in \mathcal{V}\} \right), \gamma_{\max} \left(\{\mathbf{h}_i^*, \forall v_i \in \mathcal{V}\} \right) \right],
\tag{3.10}
$$

where $\gamma_{mean}(\cdot)$ and $\gamma_{max}(\cdot)$ are the mean and max pooling operations, respectively. Ultimately, an MLP with two layers is empirically chosen as the final compatibility estimation, in which the outfit embedding is fed to compute the final compatibility score of the outfit $\mathcal{O}$ as follows,

$$\hat{y} = \sigma\left(\mathbf{W}_2\left(\alpha\left(\mathbf{W}_1\tilde{\mathbf{h}} + \mathbf{b}_1\right)\right) + \mathbf{b}_2\right), \tag{3.11}$$

where $\mathbf{W}_1 \in \mathbb{R}^{d_o \times d'}$, $\mathbf{W}_2 \in \mathbb{R}^{d' \times 1}$, $\mathbf{b}_1 \in \mathbb{R}^{d'}$ and $\mathbf{b}_2 \in \mathbb{R}^1$ are the parameters of the MLP, where d_o is the dimension of $\tilde{\mathbf{h}}$ and d' is the number of hidden units of the MLP. σ is the sigmoid active function, used for projecting the estimated compatibility score into the range of $[0, 1]$, and the estimated compatibility score can be regarded as the probability that the outfit is compatible.

Optimization To optimize the proposed model, we adopt the binary cross-entropy loss, which shows the great superiority in the classification task [28, 29], formally,

$$\mathcal{L}_{clf} = -y\log\left(\hat{y}\right) - (1 - y)\log\left(1 - \hat{y}\right), \tag{3.12}$$

where $\hat{y}$ and y denote the estimated score and the ground-truth label, respectively. Inspired by [4], to encourage the CNN to encode normalized representations in the latent space, we add additional loss to penalize the training process as follows,

$$\mathcal{L}_{emb} = \sum_{i=1}^{S} \|\mathbf{f}_i\|_2, \tag{3.13}$$

where S represents the number of fashion items in an outfit sample, and $\|\cdot\|_2$ denotes the Euclidean norm of a vector. Ultimately, the final objective function can be formulated as follows,

$$\mathcal{L}_{total} = \sum_{\Omega}\left(\mathcal{L}_{clf} + \lambda_1\mathcal{L}_{emb}\right) + \lambda_2\|\Theta\|_F^2, \tag{3.14}$$

where λ_1 and λ_2 are the tradeoff hyperparameters, controlling the weights for the normalization loss and overfitting regularization loss. As previously mentioned, Ω is the training set, and Θ refers to the set of to-be-learned parameters. $\|\cdot\|_F$ denotes the Frobenius norm of a matrix.

3.4 Experiment

To evaluate the proposed method, we conducted extensive experiments on the real-word dataset by answering the following research questions:

- RQ1: Does MOCM-MGL achieve better performance than state-of-the-art methods?
- RQ2: How does each component affect the MOCM-MGL?

- RQ3: How does each modality influence the performance?
- RQ4: How about the sensitivity of MOCM-MGL with respect to certain vital hyperparameters?

3.4.1 Experimental Settings

Dataset and Evaluation Metrics

For evaluation, similar to previous studies, we also adopted the Polyvore Outfits dataset, which has two versions: Polyvore Outfits-ND and Polyvore Outfits-D. The detailed description of this dataset was given in Chap. 2. In this work, we jointly utilized the visual images, textual descriptions and category information of each fashion item. In total, there are 11 coarse-grained categories and 154 fine-grained categories in the Polyvore Outfits dataset. Based on this dataset, we evaluated different methods on two widely recognized tasks: outfit compatibility estimation and fill-in-the-blank (FITB) fashion recommendation, both of which were detailed in Chap. 2. For these two tasks, we used the area under the receiver operating characteristic curve (AUC) and the accuracy (ACC) as the evaluation metrics, respectively.

Implementation Details

For the image encoder, we selected ResNet-18 [30] as the backbone network, and used the output of its final 4 convolutional layers (i.e., conv2_x, conv3_x, conv4_x, and conv5_x) to derive the multilayer visual representation according to Eq. (2). In this case, $L = 4$ in Eq. (2). Regarding the text encoder, we first employed the pretrained word2vec tool to obtain the 300-D vector for each word, and then fed the concatenation of all word vectors into the TextCNN. In particular, the TextCNN was equipped with $100 * 5$ filters in 3 distinct sizes [2, 3, 4]. Ultimately, we captured a 300-D textual representation for each item. As for the category encoder, we set the dimension of the category vector as 256. We empirically used the fine-gained category information. Accordingly, we set the number of category embeddings N_c as 154.

For the optimization, we employed the adaptive moment estimation method (Adam) [31]. We adopted the grid search strategy to determine the optimal values for the hyperparameters (i.e., λ_1 and λ_2) among the values $\{5e^r \mid r \epsilon - 5, \ldots, -1\}$. In addition, the learning rate, batch size, the number of propagation steps K and the dimension of the hidden state d for all methods were searched in $[1e^{-3}, 5e^{-4}, 1e^{-4}, 5e^{-5}, 1e^{-5}]$, [24, 32, 64, 128, 256], [1, 2, 3, 4, 5] and [16, 32, 64, 128, 256], respectively. The proposed model was fine-tuned based on training set and validation set for 15 epochs, and the performance on testing set was reported. We experimentally found that the model achieves the optimal performance with the initial learning rate is $5e^{-5}$ and decays by a factor of 0.5 every 10 epochs, the batch size of 32, the number of propagation steps $K = 4$, and the dimension of the hidden state

$d = 64$, respectively. The hyperparameters λ_1 and λ_2 in the loss function are $5e^{-3}$ and $5e^{-5}$, respectively. All experiments are implemented by PyTorch.

3.4.2 Model Comparison

To validate the effectiveness of our MOCM-MGL, we chose the following state-of-the-art methods as baselines.

- **Bi-LSTM** [14]: By viewing an outfit as a sequence, this method exploits the latent item interaction by the bidirectional LSTM and utilizes the VSE to capture the intermodal consistency.
- **Concatenation-Visual** [11]: This method concatenates the visual features of all fashion items into a vector and then uses an MLP as the binary classifier to compute the outfit compatibility score.
- **Concatenation-All**: For a fair comparison, this method concatenates the visual, textual and category features of all fashion items into a vector and then uses an MLP to estimate the outfit compatibility. The encoders are the same as our proposed model.
- **Pooling** [1]: This is an early fusion-based method that first concatenates the visual, textual and category features of each fashion item, and then applies the average pooling operation to aggregate fashion items.
- **Type-Aware** [3]: This method maps the item pairs into the category-specific embedding spaces and estimates the outfit compatibility by averaging all distances of the item pairs in the spaces.
- **SCE-Net** [17]: Different from Type-Aware, this method learns condition-aware embeddings by an item's characteristics without explicit category supervision. In particular, this method also uses the early fusion strategy, which integrates the visual and textual features of fashion items by the elementwise product operation.
- **NGNN** [7]: This method constructs a subgraph for each outfit, where each node represents a category and edges represent interactions among nodes. In this way, the item representation can be enhanced by that of the items in the same outfit. The outfit compatibility is jointly modeled from two channels of NGNN, whose inputs are the visual and textual modalities.
- **ABF** [20]: This is an attention-based fusion method that utilizes the attention mechanism to fuse the visual and textual features of fashion items. Since the experiment setting in [20] is consistent with ours, we directly cited the results.

Notably, all methods use the ResNet-18 as the backbone network for a fair comparison. Table 3.1 shows the performance comparison among different approaches on the Polyvore Outfits and Polyvore Outfits-D datasets under different tasks. From this table, the following observations can be made. (1) MOCM-MGL surpasses all the baselines by a large margin

Table 3.1 Performance comparison on Polyvore Outfits and Polyvore Outfits-D

Method	Polyvore Outfits		Polyvore Outfits-D	
	AUC(%)	ACC(%)	AUC(%)	ACC(%)
Bi-LSTM	66.24	38.11	62.72	37.43
Concatenation-Visual	85.21	49.93	78.62	43.05
Concatenation-All	87.61	51.35	80.23	45.14
Pooling	89.09	56.58	83.99	51.37
Type-Aware	87.23	57.78	84.49	55.85
SCE-Net	87.09	57.80	84.22	55.44
NGNN	87.12	51.79	83.61	48.37
ABF[†]	89.99	61.90	87.48	60.78
MOCM-MGL	**93.26**	**63.26**	**90.79**	**61.05**

† Indicates the results are cited from [20]

with respect to all metrics, which demonstrates the superiority of our proposed framework. (2) Concatenation-All outperforms Concatenation-Visual, which verifies the effectiveness of integrating the multimodal information of fashion items. (3) The performance of SCE-Net is similar to that of Type-Aware, demonstrating the great potential of learning condition-aware embeddings by an item characteristics instead of an explicit category indicator. (4) ABF shows superiority over all multimodal baselines, which reflects the superiority of utilizing the attention mechanism to fuse the multimodal information. And (5) it is unexpected that the graphwise method NGNN performs worse than the pairwise methods (i.e., Type-Aware and SCE-Net). The possible reason is that NGNN focuses on propagating category-oriented fashion compatibility. However, in the context of this study, the negative outfit shares the same item category as the positive outfit, which is rarely handled by NGNN.

3.4.3   Ablation Study

To explore the contribution of each component of the proposed model, we introduced following derivatives from the model.

- **w/o-MGL**: To explore the effect of the proposed modality-oriented graph learning scheme, we disabled the module by directly concatenating the visual, textual and category features of each fashion item obtained by the multimodal embedding module, and then fed it into the outfit compatibility estimation.
- **w/o-Inter**: To validate the necessity of exploring the intermodal compatibility among fashion items, we redefined the edge representation between two item nodes as

Table 3.2 Ablation study on Polyvore Outfits and Polyvore Outfits-D datasets

Method	Polyvore Outfits		Polyvore Outfits-D	
	AUC(%)	ACC(%)	AUC(%)	ACC(%)
w/o-MGL	92.31	61.84	88.93	58.44
w/o-Inter	92.95	62.87	89.18	59.27
w/o-Edge	92.92	62.75	89.76	59.35
w/o-GRU	92.45	62.41	88.97	58.77
w/o-MultiLayer	93.07	62.53	89.57	58.94
w/o-MeanPool	93.13	62.29	90.07	60.16
w/o-MaxPool	92.72	61.87	89.08	58.12
MOCM-MGL	**93.26**	**63.26**	**90.79**	**61.05**

$\mathbf{e}_{ij}^k = \alpha \left(\mathbf{W}_e^k \left(\mathbf{W}_n^k \mathbf{h}_i^{k-1} \odot \mathbf{W}_n^k \mathbf{h}_j^{k-1} \right) + \mathbf{b}_e^k \right)$, where $\mathbf{h}_i^k = \left[\mathbf{h}_{i,1}^k, \mathbf{h}_{i,2}^k, \mathbf{h}_{i,3}^k \right]$. In this way, only the intramodal compatibility was considered.

- **w/o-Edge**: To investigate the importance of edge-based compatibility relation modeling, we removed the edge representation generation unit and directly aggregated information from the hidden neighbor states, i.e., we changed Eq. (3.6) to $\mathbf{m}_{\mathcal{N}(i),pq}^k = \text{AGG} \left(\left\{ \mathbf{h}_{j,q}^k, \forall j \in \mathcal{N}(i) \right\} \right)$.
- **w/o-GRU**: To verify whether it is necessary to retain the original hidden information of the node when updating the node representation, we removed the GRU unit and only utilized the aggregation information, i.e., we changed Eq. (3.8) to $\mathbf{h}_{i,p}^k = \mathbf{m}_{\mathcal{N}(i),p}^k$.
- **w/o-MultiLayer**: To explore the importance of integrating representations obtained at different propagation layers, we treated the representation obtained at the final K th propagation layer as the updated item representation, i.e., we made $\mathbf{h}_i^* = \mathbf{h}_i^K$ in Eq. (3.9).
- **w/o-MeanPool**: To validate the function of the mean pooling operation in the outfit compatibility estimation module, we only employed the max pooling operation to generate the final outfit embedding, i.e., we rewrote Eq. (3.10) as $\tilde{\mathbf{h}} = \gamma_{max} \left(\left\{ \mathbf{h}_i^*, \forall v_i \in \mathcal{V} \right\} \right)$.
- **w/o-MaxPool**: Similarly, we removed the max pooling operation in the outfit compatibility estimation module to learn its effect by making $\tilde{\mathbf{h}} = \gamma_{mean} \left(\left\{ \mathbf{h}_i^*, \forall v_i \in \mathcal{V} \right\} \right)$ in Eq. (3.10).

Table 3.2 shows the performance comparison between MOCM-MGL and its derivatives. From this table, we obtained the following observations:

1. Our model consistently surpasses all derivations across all metrics, demonstrating the effectiveness of each component in the proposed MOCM-MGL.

2. MOCM-MGL demonstrates superiority over w/o-MGL, which implies that the modality-oriented GCN can propagate the intramodal and intermodal compatibility relation among fashion items, and therefore boost the expressiveness of item representations.
3. MOCM-MGL outperforms w/o-Inter, implying the necessity of investigating the intermodal compatibility among fashion items, to fully explore the fine-grained compatibility relation among items.
4. MOCM-MGL achieves better performance than w/o-Edge. This confirms the benefit of the edge-based compatibility relation modeling and the compatibility relation propagation during the outfit compatibility modeling.
5. MOCM-MGL surpasses w/o-GRU, which implies that selectively absorbing the compatibility information from the nodes' neighbors and the original hidden information of the node can boost the model performance.
6. w/o-MultiLayer performs worse than our MOCM-MGL. This implies that different propagation layers absorb the neighbor compatibility information at different levels, and contribute to the comprehensive outfit compatibility estimation.
7. MOCM-MGL shows superiority over w/o-MaxPool and w/o-MeanPool. This suggests that both the most prominent and the overall hidden states of fashion items are beneficial to the outfit compatibility modeling. Additionally, we observed that w/o-MeanPool outperforms w/o-MaxPool, which reflects that the max pooling operation is more effective than the mean pooling operation. This indicates that the most prominent feature of all composing items' hidden states, compared with the overall feature, has a greater impact on the outfit compatibility estimation.

3.4.4 Modality Comparison

To investigate the influence of different modalities (i.e., visual image, textual description, and category) on the performance, we compared the MOCM-MGL with different modality combinations. Notably, due to the concern that the negative outfit shares the same item categories as the positive outfit, we did not adopt the method that only utilizes category information for comparison. In addition, there are two kinds of item categories: coarse-grained categories and fine-grained categories. Therefore, there are nine modality combinations: Visual, Visual+Category (coarse), Visual+Category (fine), Textual, Textual+Category (coarse), Textual+Category (fine), Visual+Textual, All (coarse) and All (fine), where coarse, fine and All indicate that coarse-grained categories, fine-grained categories and all the three modalities are used, respectively. Table 3.3 shows the performance of our model with the nine different modality combinations. As can be seen in Table 3.3, we observed that (1) Visual outperforms Textual. This demonstrates that the visual modality is more effective than the textual feature for the outfit compatibility modeling. (2) Multimodal Visual+Textual achieves better performance than single-modal Visual and Textual. This indicates that the visual and textual modalities of fashion items complement each other in outfit compatibility estima-

Table 3.3 The performance of our proposed method with different modality combinations

Method	Polyvore Outfits		Polyvore Outfits-D	
	AUC(%)	ACC(%)	AUC(%)	ACC(%)
Visual	90.77	57.45	85.95	52.93
Visual+Category (coarse)	90.85	59.60	85.98	54.64
Visual+Category (fine)	91.02	59.67	86.01	54.75
Textual	77.02	40.33	75.62	39.01
Textual+Category (coarse)	78.41	41.66	76.71	40.92
Textual+Category (fine)	79.75	42.11	76.95	41.15
Visual+Textual	92.55	61.05	89.17	57.55
All (coarse)	93.06	62.40	90.01	59.81
All (fine)	**93.26**	**63.26**	**90.79**	**61.05**

tion. (3) Visual+Textual performs better than Visual+Category and Textual+Category. This may be attributed to the fact that the visual image and textual description deliver more content-related features of fashion items than the category information. (4) All surpasses Visual+Textual, indicating that incorporating the category information as one essential modality does improve the model performance. (5) the methods with fine-grained categories perform better than those with coarse-grained categories. This may be due to the fact that fine-grained categories provide more detailed fashion items information, which facilitates the outfit compatibility modeling.

To gain an intuitive understanding of the impact of the multimodal integration, we showed several results obtained by MOCM-MGL with different modality combinations (i.e., Visual, Textual and All) on the FITB task in Fig. 3.4. We found that only considering a single modality of fashion items may lead to incorrect choices. For instance, in the first example, Textual chooses the wrong answer d. This may be due to the fact that the textual description of the answer d and that of the given gloves share the same color, i.e., "black". Nevertheless, further incorporating the visual modality, the method All gives the correct answer c. Regarding the third example, Visual fails to give the correct answer, while All does. This makes sense, as the textual description of the ground-truth answer c shares the same pattern with the given striped hoodie. These examples demonstrate the necessity of incorporating the multimodal information in outfit compatibility modeling.

3.4.5   Hyperparameter Discussion

In this section, we examined how the number of propagation steps K, the number of convolutional layers of ResNet-18 used for visual encoding (i.e., L in Eq. (3.2)) and the number of composing items affect the performance of our method.

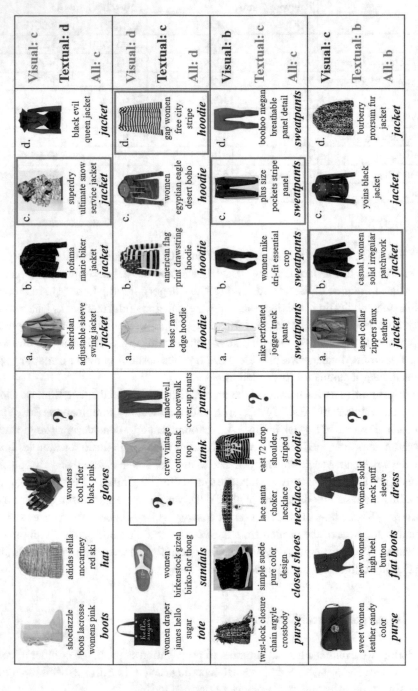

Fig. 3.4 Comparison of Visual, Textual and All on the FITB task. The black, green and red bold fonts represent the category information of fashion items, true chosen, and false chosen, respectively. The items highlighted in the green boxes are the ground truth

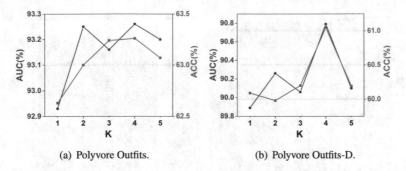

(a) Polyvore Outfits.   (b) Polyvore Outfits-D.

Fig. 3.5 Effect of the number of propagation steps, i.e., K, on Polyvore Outfits and Polyvore Outfits-D datasets

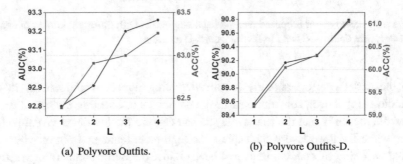

(a) Polyvore Outfits.   (b) Polyvore Outfits-D.

Fig. 3.6 Effect of the number of convolutional layers of ResNet-18 i.e., L, on Polyvore Outfits and Polyvore Outfits-D datasets

To explore the impact of the number of propagation steps, we evaluated our model's performance on two tasks with two datasets by changing K from 1 to 5 with a step of 1. As shown in Fig. 3.5, our model achieves the optimal performance when K is 4. This suggests that it is necessary to propagate several runs so that the fashion items can absorb the neighbor compatibility information thoroughly at different levels. Moreover, when K is higher than 4, the performance drops. One possible reason is that superfluous information propagation might introduce more noise into the node representations, therefore, leading to a negative effect.

We then studied the influence of the number of convolutional layers of ResNet-18 used for visual encoding, i.e., L in Eq. (3.2), on the model performance. In particular, we varied the number of convolutional layers used for visual encoding from 1 to 4. Specifically, $L = 1$ indicates that we only used the output of the final layer conv5_x of ResNet-18 for visual encoding, while $L = 2$ indicates that we used the output of the final two layers (i.e., conv4_x, and conv5_x) of ResNet-18. The cases of $L = 3$ and $L = 4$ can be similarly derived. In a sense, the larger the L, the shallower the layers' output that is incorporated. Figure 3.6 shows the performance of our model on the two tasks with the two datasets. As can be seen in

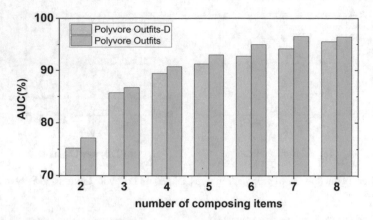

Fig. 3.7 Performance of our proposed method regarding outfits with different number of composing items on Polyvore Outfits-ND and Polyvore Outfits-D datasets

Fig. 3.6, the model's performance grows with integrating more convolutional layers' output, which indicates that each convolutional layer contributes to boosting the visual encoding. The possible reason is that the shallow layers can capture the low-level visual features of the item, while the deep layers can capture the high-level features, both of which benefit the visual encoding of fashion items and hence boost the outfit compatibility estimation performance.

To gain deeper insights, we examined the performance of our proposed model regarding outfits with different numbers of composing items. In particular, the testing set is divided according to the number of fashion items, ranging from 2 to 8. Figure 3.7 shows the performance of our proposed method with different testing configurations. As can be seen, our method performs well in all settings, verifying the effectiveness of our method to handle outfit compatibility modeling with different composing item numbers. In addition, our method performs better for outfits with multiple (i.e., more than 2) fashion items compared to those with two fashion items. This may be because the benefit of modeling the comparability between two items with a graph is limited.

3.5 Summary

In this chapter, we presented a multimodal outfit compatibility modeling scheme with modality-oriented graph learning, named MOCM-MGL, which fully exploits the visual, textual, and category modalities with GCN. Different from previous work, we treat the category information of fashion items as a unique and comparable modality to the visual and textual modalities. In addition, the proposed MOCM-MGL jointly unifies the intramodal and intermodal compatibility relation among fashion items. Extensive experiments were conducted on the Polyvore Outfits-ND and Polyvore Outfits-D datasets. The experimental results demonstrated the superiority of MOCM-MGL, suggesting that employing a modality-oriented GCN to propagate the intramodal and intermodal compatibility relation among fashion items is helpful to boost the model performance. In addition, integrating the multimodal information of fashion items greatly improves the outfit compatibility estimation performance.

References

1. Li, Yuncheng, Liangliang Cao, Jiang Zhu, and Jiebo Luo. 2017. Mining Fashion Outfit Composition Using an End-to-End Deep Learning Approach on Set Data. *IEEE Transactions on Multimedia* 19 (8): 1946–1955.
2. Dong, Xue, Jianlong Wu, Xuemeng Song, Hongjun Dai, and Liqiang Nie. 2020. Fashion Compatibility Modeling through a Multi-modal Try-on-guided Scheme. In *Proceedings of the International ACM SIGIR Conference on Research and Development in Information Retrieval*, 771–780. ACM.
3. Vasileva, Mariya I., Bryan A. Plummer, Krishna Dusad, Shreya Rajpal, Ranjitha Kumar, and David A. Forsyth. 2018. Learning Type-Aware Embeddings for Fashion Compatibility. In *European Conference on Computer Vision*, 405–421. Springer.
4. Wang, Xin, Bo Wu, and Yueqi Zhong. 2019. Outfit Compatibility Prediction and Diagnosis with Multi-layered Comparison Network. In *Proceedings of the International ACM Conference on Multimedia*, 329–337. ACM.
5. Lei Zhu, Xu Lu, Zhiyong Cheng, Jingjing Li, and Huaxiang Zhang. 2020. Deep Collaborative Multi-view Hashing for Large-scale Image Search. *IEEE Transactions on Image Processing* 29 (2020): 4643–4655.
6. Lu, Xu, Lei Zhu, Zhiyong Cheng, Liqiang Nie, and Huaxiang Zhang. 2019. Online Multimodal Hashing with Dynamic Query-adaption. In *Proceedings of the International ACM SIGIR Conference on Research and Development in Information Retrieval*, 715–724. ACM.
7. Cui, Zeyu, Zekun Li, Shu Wu, Xiaoyu Zhang, and Liang Wang. 2019. Dressing as a Whole: Outfit Compatibility Learning Based on Node-wise Graph Neural Networks. In *Porceedings of the World Wide Web Conference*, 307–317. ACM.
8. Li, Xingchen, Xiang Wang, Xiangnan He, Long Chen, Jun Xiao, and Tat-Seng Chua. 2020. Hierarchical Fashion Graph Network for Personalized Outfit Recommendation. In *Proceedings of the international ACM SIGIR Conference on Research and Development in Information Retrieval*, 159–168. ACM.
9. Yoon Kim. 2014. Convolutional Neural Networks for Sentence Classification. In *Proceedings of the Conference on Empirical Methods in Natural Language Processing*, 1746–1751. ACL.
10. Gardner, Matt W., and S.R. Dorling. 1998. Artificial Neural Networks (The Multilayer Perceptron)-A Review of Applications in the Atmospheric Sciences. *Atmospheric Environment* 32 (14–15): 2627–2636.
11. Tangseng, Pongsate, Kota Yamaguchi, and Takayuki Okatani. 2018. Recommending Outfits from Personal Closet. In *IEEE Winter Conference on Applications of Computer Vision*, 269–277. IEEE.
12. Cucurull, Guillem, Perouz Taslakian, and David Vázquez. 2019. Context-Aware Visual Compatibility Prediction. In *Proceedings of the IEEE Conference on Computer Vision and Pattern Recognition*, 12617–12626. IEEE.

13. Chaidaroon, Suthee, Yi Fang, Min Xie, and Alessandro Magnani. 2019. Neural Compatibility Ranking for Text-based Fashion Matching. In *Proceedings of the International ACM SIGIR Conference on Research and Development in Information Retrieval*, 1229–1232. ACM.
14. Han, Xintong, Zuxuan Wu, Yu-Gang Jiang, and Larry S. Davis. 2017. Learning Fashion Compatibility with Bidirectional LSTMs. In *Proceedings of the ACM International Conference on Multimedia*, 1078–1086. ACM.
15. Kiros, Ryan, Ruslan Salakhutdinov, and Richard S Zemel. 2014. Unifying Visual-semantic Embeddings with Multimodal Neural Language Models. arXiv:1411.2539.
16. Wei, Yinwei, Xiang Wang, Weili Guan, Liqiang Nie, Zhouchen Lin, and Baoquan Chen. 2020. Neural Multimodal Cooperative Learning Toward Micro-video Understanding. *IEEE Transactions on Image Processing* 29: 1–14.
17. Tan, Reuben, Mariya I. Vasileva, Kate Saenko, and Bryan A. Plummer. 2019. Learning Similarity Conditions Without Explicit Supervision. In *Proceedings of the IEEE International Conference on Computer Vision*, 10372–10381. IEEE.
18. Yang, Xun, Yunshan Ma, Lizi Liao, Meng Wang, and Tat-Seng Chua. 2019. TransNFCM: Translation-Based Neural Fashion Compatibility Modeling. In *AAAI Conference on Artificial Intelligence*, 403–410. AAAI Press.
19. Sun, Guang-Lu, Jun-Yan. He, Wu. Xiao, Bo. Zhao, and Qiang Peng. 2020. Learning Fashion Compatibility Across Categories with Deep Multimodal Neural Networks. *Neurocomputing* 395: 237–246.
20. Laenen, Katrien, and Marie-Francine. Moens. 2020. A Comparative Study of Outfit Recommendation Methods with a Focus on Attention-based Fusion. *Information Processing and Management* 57 (6): 102316.
21. Lu, Zhi, Yang Hu, Yunchao Jiang, Yan Chen, and Bing Zeng. 2019. Learning Binary Code for Personalized Fashion Recommendation. In *IEEE Conference on Computer Vision and Pattern Recognition*, 10562–10570. IEEE.
22. Kollias, Dimitrios, and Stefanos P. Zafeiriou. 2020. Exploiting Multi-CNN Features in CNN-RNN based Dimensional Emotion Recognition on the OMG in-the-wild Dataset. *IEEE Transactions on Affective Computing*, 1–12.
23. Yang, Xin, Xuemeng Song, Xianjing Han, Haokun Wen, Jie Nie, and Liqiang Nie. 2020. Generative Attribute Manipulation Scheme for Flexible Fashion Search. In *Proceedings of the International ACM SIGIR Conference on Research and Development in Information Retrieval*, 941–950. ACM.
24. Severyn, Aliaksei, and Alessandro Moschitti. 2015. Twitter Sentiment Analysis with Deep Convolutional Neural Networks. In *Proceedings of the International ACM SIGIR Conference on Research and Development in Information Retrieval*, 959–962. ACM.
25. Guo, Bao, Chunxia Zhang, Junmin Liu, and Xiaoyi Ma. 2019. Improving Text Classification with Weighted Word Embeddings via a Multi-channel TextCNN Model. *Neurocomputing* 363 (21): 366–374.
26. Mikolov, Tomás, Kai Chen, Greg Corrado, and Jeffrey Dean. 2013. Efficient Estimation of Word Representations in Vector Space. In *Proceedings of the International Conference on Learning Representations*, 1–12. OpenReview.net.
27. Li, Yujia, Daniel Tarlow, Marc Brockschmidt, and Richard S. Zemel. 2016. Gated Graph Sequence Neural Networks. In *International Conference on Learning Representations*, 15–25. OpenReview.net.
28. Guo, Sheng, Weilin Huang, Xiao Zhang, Prasanna Srikhanta, Yin Cui, Yuan Li, Hartwig Adam, Matthew R. Scott, and Serge Belongie. 2019. The iMaterialist Fashion Attribute Dataset. In *Proceedings of International Conference on Computer Vision Workshops*, 3113–3116. IEEE.

29. Inoue, Naoto, Edgar Simo-Serra, Toshihiko Yamasaki, and Hiroshi Ishikawa. 2017. Multi-Label Fashion Image Classification With Minimal Human Supervision. In *Proceeding of IEEE International Conference on Computer Vision Workshops*, 2261–2267. IEEE.
30. He,Kaiming, Xiangyu Zhang, Shaoqing Ren, and Jian Sun. 2016. Deep Residual Learning for Image Recognition. In *Proceedings of the IEEE Conference on Computer Vision and Pattern Recognition*, 770–778. IEEE Computer Society.
31. Kingma, Diederik P., and Jimmy Ba. 2015. Adam: A Method for Stochastic Optimization. In *Proceedings of the International Conference on Learning Representations*, 1–15. OpenReview.net.

Unsupervised Disentangled Graph Learning for OCM

4

4.1 Introduction

In Chaps. 2 and 3, we proposed two methods for outfit compatibility modeling. They still suffer from two key limitations. (1) They evaluate the outfit compatibility based on the single latent compatibility space. The outfit compatibility is essentially affected by multiple complementary hidden factors, such as the color, style, shape, and material. Therefore, we argue that previous methods can only achieve the suboptimal solution, as it entangles all the factors in a single latent space. (2) They focus on learning the representation of each composing item and based on that, calculate the outfit compatibility. We argue that this method still fails to authentically treat the outfit as a whole, namely, it overlooks the global outfit representation learning. Therefore, in this chapter, we aim to estimate the compatibility of the outfit by considering the multiple hidden spaces and the global outfit graph representation learning.

However, this is a nontrivial task due to the following challenges. (1) The key to outfit compatibility modeling is to learn the global outfit representation that encodes the outfit's compatibility. As the global outfit representation cannot be discussed without the local item representation learning, how to derive the accurate item representation that compiles its compatibility to all the other items poses the first challenge for us. (2) Since each outfit involves a variable number of composing items, and different items contribute to the outfit differently, how to adaptively learn the global outfit representation based on the item representation is a crucial challenge. (3) The hidden factors complementarily characterize the outfit compatibility, such as the color-oriented, material-oriented, and style-oriented compatibility. Therefore, how to model the complementarity of these hidden factors and boost the outfit compatibility modeling constitutes another difficult challenge.

To address these challenges, we devise a novel outfit compatibility modeling scheme, termed OCM-CF. As shown in Fig. 4.1, OCM-CF contains two essential components:

© The Author(s), under exclusive license to Springer Nature Switzerland AG 2022
W. Guan et al., *Graph Learning for Fashion Compatibility Modeling*,
Synthesis Lectures on Information Concepts, Retrieval, and Services,
https://doi.org/10.1007/978-3-031-18817-6_4

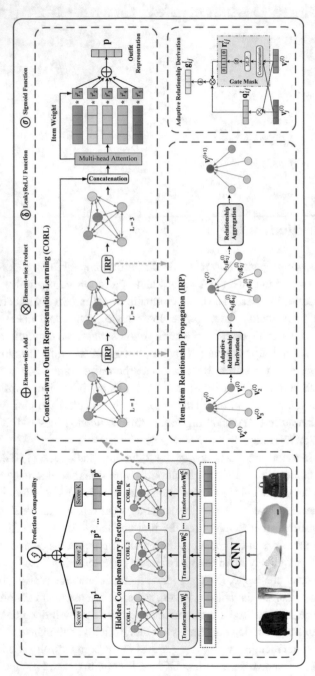

Fig. 4.1 Illustration of the proposed OCM-CF. **Left**: the overall scheme that employs a set of K parallel branches for the hidden complementary factors learning, where each branch corresponds to a factor-oriented context-aware outfit representation learning (CORL). **Right**: the detailed CORL component, and its adaptive item-item relationship propagation module

context-aware outfit representation modeling and *hidden complementary factors modeling*. Specifically, the context-aware outfit representation modeling focuses on learning the global representation of the outfit. In particular, we adopt graph convolutional networks (GCNs) to flexibly support the compatibility modeling for the outfit with an arbitrary number of fashion items. During information propagation, different from existing studies that only propagate the item embedding, we focus on propagating the item-item relationship and propose an adaptive item-item relationship propagation module based on the gate mechanism. In addition, to derive the global outfit representation, we employ the multihead attention mechanism to encourage the global outfit representation to fully incorporate the context information of each fashion item. Pertaining to the hidden complementary factors modeling, we introduce a few parallel branches, each of which is deployed with the network of the first component, i.e., context-aware outfit representation modeling, and works on exploring the outfit compatibility on one exclusive complementary hidden factor. To encourage each branch to concentrate on learning one aspect and make the whole scheme comprehensive, we introduce the orthogonality-based complementarity regularization to avoid the factor homogenization.

4.2 Related Work

Due to the remarkable capability of dealing with the unstructured data, like a graph, GNNs have been adopted in many research domains, such as the node classification [1, 2], image retrieval [3, 4], and personalized recommendation [5, 6]. Initially, Gori et al. [7] proposed GNNs to model a set of items and their relationship. Later, GCNs [1, 8] were devised to introduce the convolution operation into the graph domain by updating each node's representation via aggregating information from its neighbor nodes. To improve the model generalization ability, Velickovic et al. [9] devised a graph attention network, which assigns different importance to different neighbor nodes during graph propagation, while Hamilton et al. [10] proposed a general inductive framework that can leverage node features to efficiently generate node embeddings for unseen data by learning aggregator functions. Inspired by the success of these studies, in this work, we employed GCNs to support the compatibility modeling for the outfit with a variable length, where we developed an adaptive item-item relationship propagation module based on the gate mechanism to promote the outfit compatibility modeling performance.

4.3 Methodology

In this section, we first formally define the research task and then detail the proposed OCM-CF.

4.3.1 Problem Formulation

Formally, assume we have a set of positive (well-composed) outfits $\mathcal{S} = \{s^1, s^2, \ldots, s^T\}$ and a set of fashion items $\mathcal{X}$. Each outfit is associated with a set of m fashion items, denoted as $s = \{x_1, x_2, \ldots, x_m\}$, where x_j is the jth item of the outfit. Notably, m is a variable, which differs for different outfits. Each item x_j has a product image denoted as I_j and a category metadata denoted as $C_v \in \mathcal{C}, v \in \{1, 2, \ldots, N_c\}$, where $\mathcal{C} = \{C_1, C_2, \ldots, C_{N_c}\}$ refers to the whole set of N_c categories used for organizing all the fashion items.

In this work, we aim to devise an outfit compatibility modeling network $\mathcal{F}$, which can assess the overall compatibility score of a given outfit s as follows,

$$\hat{y} = \mathcal{F}(\{x_j\}_{j=1}^m | \Theta_F), \tag{4.1}$$

where $\hat{y}$ denotes the estimated compatibility score of the given outfit and Θ_F is a set of to-be-learned model parameters.

4.3.2 Context-Aware Outfit Representation Learning

We argue that the essence of outfit compatibility modeling is to learn a precise outfit representation that captures the compatibility among all its composition items. Due to the remarkable performance of GCNs in unstructured data representation learning, we employ GCNs to handle the outfit representation learning.

Item Visual Embedding. To begin, we first extract the image feature via the convolutional neural network (CNN) model, which can be defined as follows,

$$\mathbf{f}_j = \mathbf{CNN}(I_j; \Theta_{cnn}), \tag{4.2}$$

where $\mathbf{f}_j \in \mathbb{R}^d$ denotes the image embedding of the item x_j, d is the embedding size, and Θ_{cnn} refers to the parameters of the **CNN** model. Specifically, following [11, 12], we adopt a 18-layer deep residual network [13] pretrained on ImageNet [14]. To alleviate the overfitting, we use the L_2 regularization on the learned image embedding [15], as follows,

$$\mathcal{L}_2(s) = \sum_{j=1}^m \|\mathbf{f}_j\|_2. \tag{4.3}$$

Outfit Graph Construction. Formally, the graph for the outfit s can be defined as $\mathcal{G} = (\mathcal{V}, \mathcal{E})$, where $\mathcal{V} = \{v_1, v_2, \ldots, v_m\}$ refers to the set of item nodes, while $\mathcal{E} = \{(v_i, v_j, e_{ij}) | i \neq j, v_i \in \mathcal{V}, v_j \in \mathcal{V}\}$ denotes the set of edges linking these item nodes. The triplet (v_i, v_j, e_{ij}) denotes the edge from the node v_i to node v_j weighted by e_{ij}. Regarding the node representation initialization, since the visual cue is essential for the compatibility reasoning, we initialize each node embedding, denoted as $\mathbf{v}_j^0, j = 1, 2, \ldots, m$, with the corresponding item's visual embedding, i.e., $\mathbf{v}_j^0 = \mathbf{f}_j$. Pertaining to the edge weight, instead of setting all the edge weights as the constant, we resort to the category co-occurrence probability, due to the concern that an item should address those items whose categories frequently co-occurred with its own category to attentively absorb the neighbors' information. For example, according to the category occurrence derived from our dataset, a T-shirt should attend to the pants more as compared to the pair of glasses in the same outfit.

Thus, we introduce the category correlation matrix $\mathbf{M} \in \mathbb{R}^{N_c \times N_c}$ in a data-driven manner, which is defined as follows,

$$\begin{cases} P(C_u|C_v) = \dfrac{n_1(C_u, C_v)}{n_2(C_v)}, \\ \mathbf{M}_{uv} = \dfrac{P(C_u|C_v)}{\sum_{k=1}^{N_c} P(C_u|C_k)}, \end{cases} \tag{4.4}$$

where $P(C_u|C_v)$ denotes the occurrence probability of category C_u given category C_v. $n_1(C_u, C_v)$ is the function for counting the concurrence times of categories C_u and C_v in the training dataset, and $n_2(C_v)$ is that for counting the occurrence times of category C_v in the training dataset. Assume that items x_i and x_j belong to the categories C_u and C_v, respectively. Then we define the weight for the edge from x_i to x_j as,

$$e_{ij} = \mathbf{M}_{uv}. \tag{4.5}$$

Item-Item Relationship Propagation (IRP). Different from existing work [16, 17] that propagates the pure neighbor items' embedding over the item graph, we propose to propagate the item-item relationship embedding, a.k.a., adaptive relationship derivation, which plays a pivotal role in outfit compatibility modeling. Additionally, we argue that the high-order connectivities arc beneficial to synthesizing a richer node representation [5, 6], and thus stack L propagation layers to exploit the item-item relationship. Specifically, we define the relationship embedding between the item i and j regarding the lth propagation layer as $\mathbf{q}_{ij}^l = \mathbf{v}_i^l \otimes \mathbf{v}_j^l$, where $\otimes$ denotes the element-wise product operation, and $l = 1, 2, \ldots, L$.

Regarding the item-item relationship propagation, we argue that different dimensions of the relationship embedding may contribute differently to the compatibility modeling. Accordingly, we introduce the gate mechanism to adaptively propagate the item-item relationship. In particular, the gate function is defined as follows,

$$\mathbf{r}_{ij}^l = \sigma\left(\mathbf{W}_1^l \delta\left(\mathbf{W}_2^l(\mathbf{v}_i^l || \mathbf{v}_j^l) + \mathbf{b}_2^l\right) + \mathbf{b}_1^l\right), \tag{4.6}$$

where $\mathbf{r}_{ij}^l \in \mathbb{R}^d$ is the gate mask for the item pair (x_i, x_j) in the lth propagation layer, $||$ is the concatenation operation, $\sigma(\cdot)$ and $\delta(\cdot)$ are Sigmoid and LeakyReLU [18] activate functions, respectively. $\mathbf{W}_1^l \in \mathbb{R}^{d \times d}$, $\mathbf{W}_2^l \in \mathbb{R}^{d \times 2d}$, $\mathbf{b}_1^l \in \mathbb{R}^d$, and $\mathbf{b}_2^l \in \mathbb{R}^d$ are trainable parameters of the fully connected layers for the lth order relationship propagation. Based upon the gate function, we formulate the final item-item relationship as $\mathbf{g}_{ij}^l = \delta(\mathbf{r}_{ij}^l \otimes \mathbf{q}_{ij}^l)$.

Thereafter, we aggregate all the neighbor relationships to refine the ego item representation. Mathematically, the item-item relationship propagation for item j in the lth order propagation can be formulated as,

$$\mathbf{v}_j^{(l+1)} = \delta\left(\mathbf{W}_3^l\left(\mathbf{f}_j^l + \sum_{i \in \mathcal{N}_j} e_{ij}\mathbf{g}_{ij}^l\right) + \mathbf{b}_3^l\right), \tag{4.7}$$

where $\mathcal{N}_j$ is the set of neighbor nodes of the node x_j. $\mathbf{W}_3^l$ and $\mathbf{b}_3^l$ are learnable parameters for node information aggregation in the lth propagation layer. Finally, to avoid the information loss during the item-item relationship propagation, we incorporate the initial visual embedding to define the final item embedding as follows,

$$\hat{\mathbf{v}}_j = \mathbf{v}_j^0 \| \mathbf{v}_j^L, \tag{4.8}$$

where $\hat{\mathbf{v}}_j \in \mathbb{R}^{2d}$ is the final representation of the item x_j.

To encourage the gate mask to filter the discriminative dimensions of the relationship, we introduce the L1 regularization to enhance the sparsity of gate masks as follows,

$$\mathcal{L}_1(s) = \sum_{l=1}^{L}\sum_{i=1}^{m}\sum_{j=1,j\neq i}^{m} \|\mathbf{r}_{ij}^l\|_1. \tag{4.9}$$

Global Outfit Representation.

Different from existing graph-based compatibility modeling methods [17, 19] that focus on learning the individual compatibility of each item for the outfit based on the local item representation learning, we directly target the global outfit representation learning. To discriminate the importance of different items in characterizing the outfit, we adopt the multihead self-attention mechanism [20] to summarize the outfit representation from the set of item representations.

For simplicity, we pack all item embeddings into a matrix $\widehat{\mathbf{V}}^F = [\hat{\mathbf{v}}_1; \hat{\mathbf{v}}_2; \ldots; \hat{\mathbf{v}}_m] \in \mathbb{R}^{m \times 2d}$. Assume we have h attention heads, and the self-attention function of the ith attention head can be formulated as follows,

$$\begin{cases} \mathbf{S}_i^a = softmax(\dfrac{\mathbf{Q}_i\mathbf{K}_i^{\mathbf{T}}}{\sqrt{d_k}}), \\ \mathbf{H}_i = \mathbf{S}_i^a\mathbf{Z_i}, \end{cases} \tag{4.10}$$

where $\mathbf{Q}_i = \widehat{\mathbf{V}}^F\mathbf{W}_i^Q$, $\mathbf{K}_i = \widehat{\mathbf{V}}^F\mathbf{W}_i^K$, and $\mathbf{Z}_i = \widehat{\mathbf{V}}^F\mathbf{W}_i^Z$ refer to the query, key and value matrices, respectively, while $\mathbf{W}_i^Q \in \mathbb{R}^{2d \times d_q}$, $\mathbf{W}_i^K \in \mathbb{R}^{2d \times d_k}$, and $\mathbf{W}_i^Z \in \mathbb{R}^{2d \times d_z}$ are the corresponding trainable linear projection matrices. d_q, d_k, and d_z are the dimensions of the latent space, where $d_q = d_k = d_z = \frac{2d}{h}$. The $softmax$ operation is performed for each row. $\mathbf{S}_i^a \in \mathbb{R}^{m \times m}$ denotes the attention weight matrix of the ith head, the (j, k)th entity of which reflects the importance of the kth item to the jth item. $\mathbf{H}_i \in \mathbb{R}^{m \times d_z}$ is the output of ith head, with each row referring to an item local feature.

Based upon the attention weight matrices derived from the h heads, we define the importance of the ith item as the summation of its importance to all the items in the outfit. Formally, we have,

$$\begin{cases} r_i = \sum_{j=1}^{m} \overline{\mathbf{S}}(j, i), i = 1, 2, \dots, m, \\ \tilde{r}_i = \dfrac{\exp(r_i)}{\sum_{k=1}^{m} \exp(r_k)}, \end{cases} \quad (4.11)$$

where $\overline{\mathbf{S}} = \frac{1}{h} \sum_{i=1}^{h} \mathbf{S}_i^a$, and $\tilde{r}_i$ is the normalized importance of the item i to characterize the outfit. Finally, we derive the outfit representation as follows,

$$\mathbf{p} = \sum_{i=1}^{m} \tilde{r}_i \mathbf{H}^F(i, :), \quad (4.12)$$

where $\mathbf{H}^F = [\mathbf{H}_1, \mathbf{H}_2, \dots, \mathbf{H}_h] \in \mathbb{R}^{m \times 2d}$, and $\mathbf{H}^F(i, :)$ is the ith row of $\mathbf{H}^F$, representing the ith item representation. Finally, our proposed context-aware outfit representation learning network can be summarized as,

$$\mathcal{F}_{out}(\mathcal{G}) = \mathbf{p}, \quad (4.13)$$

where $\mathcal{G}$ is the item graph of outfit s.

4.3.3 Hidden Complementary Factors Learning

As previously stated, each outfit's compatibility can be affected by multiple complementary hidden factors, like the color, style, shape, and material. Accordingly, in this part, we propose the hidden complementary factors learning method. In particular, we first project each outfit into multiple complementary factor subspaces, in which the factor-oriented compatibility can be modeled.

Thus, for each outfit s, we introduce K parallel branches, denoted as $\mathcal{B}_1, \dots, \mathcal{B}_K$, where each branch comprises a network for CORL, and focuses on one hidden factor-oriented outfit compatibility reasoning. Specifically, each branch $\mathcal{B}_k, k = 1, 2, \dots, K$ can be formulated as,

$$\mathbf{p}^k = \mathcal{F}_{out}^k(\mathcal{G}^k), \quad (4.14)$$

where $\mathbf{p}^k \in \mathbb{R}^{2d}$ denotes the global outfit representation pertaining to the kth hidden factor. $\mathcal{G}^k = \{\mathcal{V}^k, \mathcal{E}^k\}$ is the outfit graph designed for the kth branch, where $\mathcal{E}^k = \mathcal{E}$, namely, all the branches share the same graph structure. To facilitate the hidden factor learning, we initialize the node representations of different branches with different hidden representations. Specifically, we have,

$$\mathbf{f}_j^k = \mathbf{W}_b^k \mathbf{f}_j, \, k = 1, 2, \ldots, K, \tag{4.15}$$

where $\mathbf{f}_j$ is the initial visual embedding of item j, defined in Eq. (4.2). $\mathbf{W}_b^k \in \mathbb{R}^{d \times d}$ is the weight matrix for transforming the visual embedding of each item into the kth hidden factor space. $\mathbf{f}_j^k \in \mathbb{R}^d$ denotes the kth hidden factor-oriented item representation.

It is worth noting that with no constraint, the learned hidden factors tend to present the homogenization, resulting in redundant compatibility reasoning of different branches [21]. To encourage different branches to model different hidden factors, we introduce the orthogonality-based complementarity regularization. Formally, we have the following objective function,

$$\mathcal{L}_{com}(s) = \sum_{j=1}^{m} \|\mathbf{F}_j \mathbf{F}_j^{\mathrm{T}} - \mathbf{I}\|_F^2, \tag{4.16}$$

where $\mathbf{I} \in \mathbb{R}^{K \times K}$ is the identity matrix. $\mathbf{F}_j = [\mathbf{f}_j^1; \mathbf{f}_j^2; \ldots; \mathbf{f}_j^K] \in \mathbb{R}^{K \times d}$ denotes different factor embeddings of the fashion item j, and $\| \cdot \|_F$ denotes the Frobenius norm of the matrix.

4.3.4 Outfit Compatibility Modeling

Based on the hidden factor-oriented outfit representations, i.e., $\mathbf{p}^1, \mathbf{p}^2, \ldots, \mathbf{p}^K$, we employ a linear transformation to obtain the compatibility score $\hat{y}$ for the given outfit s as follows,

$$\hat{y} = \sum_{k=1}^{K} \mathbf{W}_s^k \mathbf{p}^k, \tag{4.17}$$

where $\mathbf{W}_s^k \in \mathbb{R}^{1 \times 2d}$ is the weight matrix for the branch $\mathcal{B}_k$.

Similar to existing methods [17, 22], to exploit the implicit compatibility preference among fashion items, we also adopt the Bayesian personalized ranking (BPR) [23] loss, which encourages the score of the positive outfit higher than that of the negative outfit. Accordingly, we first build the following training set $\mathcal{D} = \{(s^+, s^-)\}$, where s^+ and s^- denote the positive and negative outfit samples, respectively. s^+ is directly sampled from the positive outfit set $\mathcal{S}$, while s^- is strategically sampled. The sampling details are given in the experiment section. For each training pair (s^+, s^-), we have the following objective function,

$$\mathcal{L}_{bpr}(s^+, s^-) = -\ln \sigma(\hat{y}_{s^+} - \hat{y}_{s^-}). \tag{4.18}$$

Then the ultimate training loss can be defined as follows,

$$\begin{aligned} \min_{\Theta_F} \mathcal{L} = \sum_{(s^+, s^-) \in \mathcal{D}} \mathcal{L}_{bpr}(s^+, s^-) + \lambda_1 \big(\mathcal{L}_{com}(s^+) + \mathcal{L}_{com}(s^-)\big) \\ + \lambda_2 \big(\mathcal{L}_1(s^+) + \mathcal{L}_1(s^-)\big) + \lambda_3 \big(\mathcal{L}_2(s^+) + \mathcal{L}_2(s^-)\big), \end{aligned} \tag{4.19}$$

where λ_1, λ_2, and λ_3 are nonnegative tradeoff hyperparameters and Θ_F refers to the set of parameters (i.e., Θ_{cnn}, $\mathbf{W}_1^l$, $\mathbf{W}_2^l$, $\mathbf{W}_3^l$, $\mathbf{b}_1^l$, $\mathbf{b}_2^l$, $\mathbf{b}_3^l$, $\mathbf{W}_i^Q$, $\mathbf{W}_i^K$, $\mathbf{W}_i^Z$, $\mathbf{W}_b^k$ and $\mathbf{W}_s^k$) of the model.

4.4 Experiment

To evaluate the proposed method, we conducted extensive experiments on the two real-world datasets Polyvore Outfits and Polyvore Outfits-D via answering the following research questions:

- **RQ1:** Does OCM-CF surpass the state-of-the-art methods?
- **RQ2:** How does each component affect our OCM-CF?
- **RQ3:** What is the qualitative performance of OCM-CF?

4.4.1 Experimental Settings

Dataset and Evaluation Metrics
Similar to our previous studies introduced in the previous two chapters, we also adopted the Polyvore Outfits dataset, which has two versions: Polyvore Outfits and Polyvore Outfits-D, for evaluation. The detailed description of this dataset was given in Chap. 2. In this work, we jointly utilized the visual images, textual descriptions and category information of each fashion item. In total, there are 11 coarse-grained categories and 154 fine-grained categories in the Polyvore Outfits dataset.

Based upon this dataset, we evaluated different methods on two widely recognized tasks: outfit compatibility estimation and fill-in-the-blank (FITB) fashion recommendation, both of which were detailed in Chap. 2. For these two tasks, we used the area under the receiver operating characteristic curve (AUC) and the accuracy (ACC) as the evaluation metrics, respectively.

Implementation Details
Negative Outfit Composition. Regarding the training dataset construction, we set the ratio of positive and negative samples to 1 : 1. Considering that human cognitive learning is an easy-to-hard process, analogically, the model first learned from the easy cases, and hence adopted the following three methods to compose a negative outfit s^- for each positive outfit s^+: (1) Method 1: randomly sample $|s^+|$ items from $\mathcal{X}$ without any restriction; (2) Method 2: randomly sample $|s^+|$ items from $\mathcal{X}$ according to the item categories of s^+; and (3) Method 3: randomly choose one item of the positive outfit and replace it with a randomly sampled item of the same category. Intuitively, in the first few epochs, we used Method 1 to derive the negative samples, then Method 2, followed by Method 3 in the last few epochs.

Experimental Setting. In Polyvore dataset, each fashion item is assigned with both the coarse-grained category, such as *top*, and the fine-grained category, such as *t-shirt*. Due to the concern of the highly imbalanced data distribution with hundreds of fine-grained categories, which may degrade the model generalization performance, we resorted to the coarse-grained category metadata to derive the edge weight between every two items. The Adam optimizer was employed with minibatch size 64 and embedding size $d = 64$. The learning rate was set as $5e^{-5}$ with the exponential decay 0.985 of each epoch. We empirically set $L = 3$ as the propagation layers, and we stacked 3 layers of multihead self-attentions with $h = 8$ heads. In the hidden complementary factors learning module, the number of branches K was set to 5. We set $\lambda_2 = 5e^{-4}$ and other λ parameters in Eq. (4.19) to $5e^{-3}$. The proposed model was trained for 80 epochs, and the performance was reported on the test dataset. Notably, during the training, we used two thresholds regarding the epoch number to switch the negative outfit composition method as 10 and 40. We only used the image signal in all experiments for fair comparisons.

Evaluation Tasks and Metrics. We evaluated our proposed OCM-CF by conducting experiments on three popular tasks: the outfit compatibility prediction [12, 24], fill-in-the-blank [11, 17], and complementary fashion item retrieval [22, 25]. (1) The outfit compatibility prediction task evaluates the compatibility score of a given outfit that contains an arbitrary number of fashion items. Following existing studies [12, 24], we adopted the AUC (area under the ROC curve) [26] as the evaluation metric. (2) The task of fill-in-the-blank (FITB) selects the most compatible item from a set of item candidates for an incomplete query outfit. To prepare the data, for each positive/compatible outfit, we randomly selected an item as the target item and set the remaining items of the outfit as the query. Then we composed the target item with three other randomly selected items of the same category with the target item from the dataset as the candidate item choices. To handle the task, we composed each candidate item with query items as an outfit and used the well-trained model to compute each outfit's compatibility score. Based on that, we chose the item with the highest score as the answer, and used the accuracy as the evaluation metric. (3) The task of complementary fashion item retrieval can be seen as an extension of the FITB task. Specifically, we extended the size of the candidate item set to 500, where there was only one positive (target) item and 499 negative items of the same category. We adopted the hit rate (HR) at 5, 10 and 40 to evaluate the model performance.

4.4.2 Model Comparison

To validate the effectiveness of our proposed method, we compared it with the following state-of-the-art methods, including the pair-based, sequence-based, and graph-based models.

- **Bi-LSTM** [24] permutes all items of an outfit into a predefined order according to the item category, and cast the outfit compatibility modeling as a sequence prediction problem, where bidirectional LSTMs are used. For fairness, we removed the text information from the model.

Table 4.1 Performance comparison among different methods on three tasks. Our results are highlighted in bold

Method	Polyvore Outfits		Polyvore Outfits-D		Polyvore Outfits			Polyvore Outfits-D		
	Compat. AUC	FITB ACC	Compat. AUC	FITB Acc	HR@5	HR@10	HR@40	HR@5	HR@10	HR@40
Bi-LSTM	0.68	42.20%	0.65	40.10%	0.032	0.076	0.244	0.052	0.088	0.249
SCE-NET	0.83	52.80%	0.82	52.10%	0.079	0.143	0.340	0.076	0.129	0.334
Type-aware	0.87	56.60%	0.78	47.30%	0.108	0.165	0.372	0.040	0.072	0.236
NGNN	0.75	53.02%	0.68	42.49%	0.084	0.136	0.341	0.033	0.068	0.219
Context-aware	0.81	55.63%	0.77	50.34%	0.106	0.163	0.384	0.083	0.132	0.325
HFGN	0.84	49.90%	0.70	39.03%	0.050	0.080	0.288	0.023	0.049	0.164
OCM-CF	**0.92**	**63.62%**	**0.86**	**56.59%**	**0.145**	**0.238**	**0.502**	**0.096**	**0.158**	**0.370**
%Improv.	5.75%	12.40%	4.88%	8.62%	34.26%	44.24%	30.73%	15.66%	19.70%	10.78%

- **Type-aware** [12] measures the fashion item compatibility with type-respecting spaces rather than a single general space. We used the code provided by the authors and retrained the model with only the image cue.
- **SCE-NET** [11] learns different similarity conditions and employs a weight module to combine all different embeddings as a fashion item representation. Similar to type-aware, we removed the regularization of the text information from the author-released model.
- **NGNN** [17] maps the fashion item feature into a category space to build the item graph, where the node embedding is updated based on GRU [27] and the attention mechanism is used for summarizing the outfit compatibility score.
- **Context-aware** [16] builds a graph with all fashion items in the dataset. Each node will receive a message from its outfit and other outfits to learn the contextual item embedding. In the testing stage, we computed the compatibility score based on its embedding.
- **HFGN** [19] different from NGNN, devises a R-view attention map and a R-view score map to assess the outfit compatibility score based on GCNs over the category-oriented outfit graph.

Table 4.1 shows the performance comparison among different approaches on both Polyvore Outfits and Polyvore Outfits-D datasets under different tasks. For clarity, we divided the baselines into three groups, i.e., sequence-based, pair-based, and graph-based models. From this table, we make the following observations: (1) Compared to other baselines, Bi-LSTM achieves the worst performance on most evaluation metrics, which may be due to two facts. On the one hand, essentially, it is inappropriate to model the outfit as an ordered list of fashion items. On the other hand, this method computes the outfit compatibility score by predicting the next item with the previous outfits, which may cause cumulative error propagation. (2) Unexpectedly, the graph-based baselines, i.e., NGNN, Context-aware and HFGN, do not show superiority over the pair-based methods, i.e., SCE-NET and type-aware. The possible explanation for Context-aware is that this method learns fashion item embeddings

Table 4.2 Performance of the ablation study on Polyvore Outfit and Polyvore Outfit-D datasets

Method	Polyvore Outfits		Polyvore Outfits-D	
	Compat. AUC	FITB ACC (%)	Compat. AUC	FITB ACC (%)
OCM-CF	**0.92**	**63.62**	**0.86**	**56.59**
w/o Edge Weight	0.89	62.60	0.84	55.64
w/o Relationship	0.90	62.12	0.84	55.25
w/o Attention	0.64	51.12	0.61	34.11
w Fine-grained	0.87	61.93	0.80	54.65
w/o Complementarity	0.89	62.83	0.80	55.94

in a single space, while the pair-based method, type-aware, considers the visual similarity from different metric spaces. For NGNN and HFGN, they employ fashion item embeddings in a category space to initialize nodes, which leads to category bias, namely, the model may learn compatibility patterns at the category level, resulting in an inaccurate evaluation of outfit compatibility score. (3) Our proposed method OCM-CF consistently achieves the best performance on all tasks. It is worth noting that our method has large improvements on the complementary fashion item retrieval task w.r.t. HR@5 and HR@10, which is meaningful for the real-world application since users can quickly find the complementary fashion item fitting the outfit. The results verify the superiority of our model over the state-of-the-art methods, and the effectiveness of our contextual outfit representation learning and the hidden complementary factors learning.

4.4.3   Ablation Study

To investigate how each component affects our model, we introduced the following five variants:

- **w/o Edge Weight**. In this variant, we set all the edge weights as a constant 1.
- **w/o Relationship**. We modified the $\mathbf{g}_{ij}^{l} = \mathbf{v}_i^l$ in Eq. (4.7) for only aggregating neighbor item embeddings.
- **w/o Attention**. We replaced the multihead attention mechanism with a mean pooling operation over the representation of all the composition fashion items.
- **w Fine-grained**. We derived the edge weight with the fine-grained category co-occurrence [17, 19] rather than the coarse-grained category used in our OCM-CF.
- **w/o Complementarity**. We removed the orthogonality-based complementarity regularization from the hidden complementary factors learning.

Table 4.2 shows the performance comparison of different methods in the ablation study. From Table 4.2, we noticed that all the variants degrade the performance of our OCM-CF, which indicates the importance of each component. In particular, first, w/o edge weight performs worse than OCM-CF, implying that utilizing the category co-occurrence probability can promote the item-item relationship propagation. Second, w/ fine-grained is inferior to OCM-CF, which confirms our assertion that utilizing the fine-grained categories may involve the highly imbalanced data distribution, making the co-occurrence pattern unreliable. Third, the inferior performance of w/o relationship suggests that propagating the item-item relationship is more meaningful for the outfit compatibility modeling. Fourth, we found the performance of w/o attention significantly drops, as compared to OCM-CF, demonstrating the necessity of deriving the global outfit representation in an attentive manner. Last, w/o complementarity is also inferior to OCM-CF, reflecting the effectiveness of our proposed complementarity regularization.

4.4.4 Case Study

To gain a more intuitive understanding of our model, we conducted the case study on two tasks: similar outfit retrieval and complementary fashion item retrieval.

Similar Outfit Retrieval
To illustrate the effectiveness of the outfit representation learned by our model, we investigated the performance of our model in the task of similar outfit retrieval, which aims to retrieve similar outfits for a given query outfit. We argued that similar outfits tend to share the common prominent features. Instead of using all factor-oriented outfit representations, we particularly adopted the outfit representation corresponding to the highest compatibility score, i.e., $\mathbf{p}^{k^*}$, where $k^* = \arg\max_k \{\mathbf{W}_s^k \mathbf{p}^k |_{k=1}^K\}$, and employed the cosine similarity between the query outfit and each candidate outfit to retrieve the similar outfit for the query outfit. In the comprehensive evaluation, we studied the similar outfit retrieval task in two scenarios: (1) the candidate outfits have the same length, i.e., the same number of composition items with the query outfit, and (2) the candidate outfits have random lengths. In this part, we directly employed the test dataset of the compatibility prediction as the set of candidate outfits. Figure 4.2 illustrates the retrieval results of two testing outfits in each scenario. For the first scenario, from the left example in Fig. 4.2a, we observed that the retrieved outfits share the blue color, and the fresh style, while in the right example, we noticed that the returned outfits also possess the high similarity with the query outfit, such as the summer style, a variety of colors and patterns, and the item categories. Similar observations can be also obtained from Fig. 4.2b, where the length of the retrieved outfits is not restricted. In general, these observations reflect the effectiveness of the factor-oriented outfit representation learned by our model and the benefit of exploring the hidden complementary factors to capture the discriminative feature of the outfit.

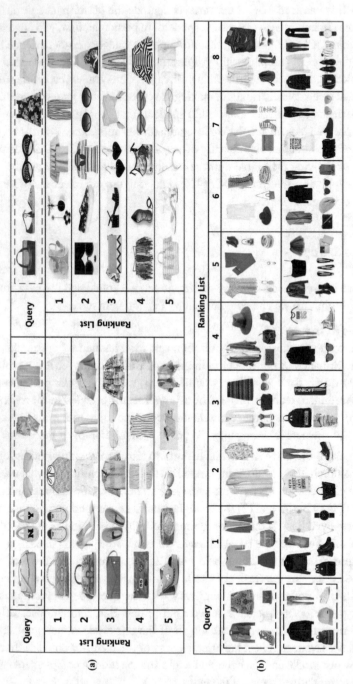

Fig. 4.2 Illustration of the similar outfit retrieval results with testing samples in two scenarios

Fig. 4.3 Illustration of the complementary item retrieval results. Positive items are highlighted in red boxes

Complementary Fashion Item Retrieval

Similar to existing studies [22, 25], we also presented the qualitative results of our model in the complementary item retrieval task, where the candidate set comprised a target item as well as nine negative items. Moreover, we adopted two negative item sampling protocols: the negative items were randomly selected from items with the same coarse-grained category with the target item, and (2) the negative items were randomly selected from items with the same fine-grained category with the target item, which corresponds to a more challenging task. In addition, we adopted the best baseline on the complementary item retrieval task, i.e., type-aware, for comparison. Due to the limited space, we only exhibited one example for each scenario in Fig. 4.3. As can be seen, our OCM-CF can rank the target items at the top places, outperforming the type-aware method.

4.5 Summary

In this chapter, we presented a novel outfit compatibility modeling scheme via complementary factorization, named OCM-CF, which seamlessly unifies the context-aware outfit representation learning and hidden complementary factors learning in the context of outfit compatibility modeling. Extensive experiments were conducted on two real-world datasets, and the encouraging experimental results validate the superiority of our proposed model and the importance of each component. In addition, we notice that the global outfit representation models one compatible factor by considering all items of the outfit with a comprehensive perspective and the proposed orthogonality-based complementarity regularization can make the factor-oriented outfit representation discriminative.

References

1. Kipf, Thomas N., and Max Welling. 2017. Semi-Supervised Classification with Graph Convolutional Networks. In *International Conference on Learning Representations*, 1–15. OpenReview.net.
2. Rong, Yu, Wenbing Huang, Tingyang Xu, and Junzhou Huang. 2020. DropEdge: Towards Deep Graph Convolutional Networks on Node Classification. In *Proceedings of the International Conference on Learning Representations*, 1–17. OpenReview.net.
3. Zhang, Zhaolong, Yuejie Zhang, Rui Feng, Tao Zhang, and Weiguo Fan. 2020. Zero-Shot Sketch-Based Image Retrieval via Graph Convolution Network. In *Proceedings of the International Joint Conference on Artificial Intelligence*, 12943–12950. AAAI Press.
4. Zhou, Xiang, Fumin Shen, Li. Liu, Wei Liu, Liqiang Nie, Yang Yang, and Heng Tao Shen. 2020. Graph Convolutional Network Hashing. *IEEE Transactions on Cybernetics* 50 (4): 1460–1472.
5. Wei, Yinwei, Xiang Wang, Liqiang Nie, Xiangnan He, Richang Hong, and Tat-Seng Chua. 2019. MMGCN: Multi-modal Graph Convolution Network for Personalized Recommendation of Micro-video. In *Proceedings of the ACM International Conference on Multimedia*, 1437–1445. ACM.
6. He, Xiangnan, Kuan Deng, Xiang Wang, Yan Li, Yong-Dong Zhang, and Meng Wang. 2020. LightGCN: Simplifying and Powering Graph Convolution Network for Recommendation. In *Proceedings of the International ACM SIGIR Conference on Research and Development in Information Retrieval*, 639–648. ACM.
7. Gori, M., G. Monfardini, and F. Scarselli. 2005. A New Model for Learning in Graph Domains. In *Proceedings of the IEEE International Joint Conference on Neural Networks*, vol. 2, 729–734.
8. Kipf, Thomas, and M. Welling. 2016. Variational Graph Auto-Encoders. In *NIPS Workshop on Bayesian Deep Learning*.
9. Veličković, Petar, Guillem Cucurull, Arantxa Casanova, Adriana Romero, Pietro Liò, and Yoshua Bengio. 2018. Graph Attention Networks. In *Proceedings of the International Conference on Learning Representations*, 45–61. OpenReview.net.
10. Hamilton, William L., Zhitao Ying, and Jure Leskovec. 2017. Inductive Representation Learning on Large Graphs. In *Advances in Neural Information Processing Systems*, 1024–1034. Curran Associates Inc.
11. Tan, Reuben, Mariya I. Vasileva, Kate Saenko, and Bryan A. Plummer. 2019. Learning Similarity Conditions Without Explicit Supervision. In *Proceedings of the IEEE International Conference on Computer Vision*, 10372–10381. IEEE.
12. Vasileva, Mariya I., Bryan A. Plummer, Krishna Dusad, Shreya Rajpal, Ranjitha Kumar, and David A. Forsyth. 2018. Learning Type-Aware Embeddings for Fashion Compatibility. In *European Conference on Computer Vision*, 405–421. Springer.
13. He, Kaiming, Xiangyu Zhang, Shaoqing Ren, and Jian Sun. 2016. Deep Residual Learning for Image Recognition. In *Proceedings of the IEEE Conference on Computer Vision and Pattern Recognition*, 770–778. IEEE Computer Society.
14. Olga Russakovsky, J., H. Deng, J. Su, S. Krause, S.. Ma. Satheesh, A. Zhiheng Huang, A. Khosla. Karpathy, Michael S. Bernstein, A. Berg, and Li. Fei-Fei. 2015. ImageNet Large Scale Visual Recognition Challenge. *International Journal of Computer Vision* 115: 211–252.

15. Veit, Andreas, Serge J. Belongie, and Theofanis Karaletsos. 2017. Conditional Similarity Networks. In *Proceedings of the IEEE Conference on Computer Vision and Pattern Recognition*, 1781–1789. IEEE.
16. Cucurull, Guillem, Perouz Taslakian, and David Vázquez. 2019. Context-Aware Visual Compatibility Prediction. In *Proceedings of the IEEE Conference on Computer Vision and Pattern Recognition*, 12617–12626. IEEE.
17. Cui, Zeyu, Zekun Li, Shu Wu, Xiaoyu Zhang, and Liang Wang. 2019. Dressing as a Whole: Outfit Compatibility Learning Based on Node-wise Graph Neural Networks. In *Porceedings of the World Wide Web Conference*, 307–317. ACM.
18. Maas, Andrew L, Awni Y Hannun, and Andrew Y Ng. 2013. Rectifier Nonlinearities Improve Neural Network Acoustic Models. In *Proceedings of the International Conference on Machine Learning*, 3–3. JMLR.org.
19. Li, Xingchen, Xiang Wang, Xiangnan He, Long Chen, Jun Xiao, and Tat-Seng Chua. 2020. Hierarchical Fashion Graph Network for Personalized Outfit Recommendation. In *Proceedings of the International ACM SIGIR Conference on Research and Development in Information Retrieval*, 159–168. ACM.
20. Vaswani, Ashish, Noam Shazeer, Niki Parmar, Jakob Uszkoreit, Llion Jones, Aidan N. Gomez, Lukasz Kaiser, and Illia Polosukhin. 2017. Attention is All you Need. In *Proceedings of the Advances in Neural Information Processing Systems*, 5998–6008. NIPS.
21. Wang, Xiang, Hongye Jin, An Zhang, Xiangnan He, Tong Xu, and Tat-Seng Chua. 2020. Disentangled Graph Collaborative Filtering. In *Proceedings of the International ACM SIGIR Conference on Research and Development in Information Retrieval*, 1001–1010. ACM.
22. Song, Xuemeng, Xianjing Han, Yunkai Li, Jingyuan Chen, Xin-Shun Xu, and Liqiang Nie. 2019. GP-BPR: Personalized Compatibility Modeling for Clothing Matching. In *Proceedings of the ACM International Conference on Multimedia*, 320–328. ACM.
23. Rendle, Steffen, Christoph Freudenthaler, Zeno Gantner, and Lars Schmidt-Thieme. 2009. BPR: Bayesian Personalized Ranking from Implicit Feedback. In *Proceedings of the International Conference on Uncertainty in Artificial Intelligence*, 452–461. AUAI Press.
24. Han, Xintong, Zuxuan Wu, Yu-Gang Jiang, and Larry S. Davis. 2017. Learning Fashion Compatibility with Bidirectional LSTMs. In *Proceedings of the ACM International Conference on Multimedia*, 1078–1086. ACM.
25. Dong, Xue, Jianlong Wu, Xuemeng Song, Hongjun Dai, and Liqiang Nie. 2020. Fashion Compatibility Modeling through a Multi-modal Try-on-guided Scheme. In *Proceedings of the International ACM SIGIR Conference on Research and Development in Information Retrieval*, 771–780. ACM.
26. Zhang, Hanwang, Zheng-Jun Zha, Yang Yang, Shuicheng Yan, Yue Gao, and Tat-Seng Chua. 2013. Attribute-Augmented Semantic Hierarchy: Towards Bridging Semantic Gap and Intention Gap in Image Retrieval. In *Proceedings of the ACM International Conference on Multimedia*, 33–42. ACM.
27. Li, Yujia, Daniel Tarlow, Marc Brockschmidt, and Richard S. Zemel. 2016. Gated Graph Sequence Neural Networks. In *International Conference on Learning Representations*, 15–25. OpenReview.net.

5.1 Introduction

In Chap. 4, we studied the fine-grained outfit compatibility modeling, where the hidden factors affecting the outfit compatibility are jointly considered. One key limitation is that it only investigates the visual content of fashion items while overlooking the items' semantic attributes. The item attribute labels usually contain rich information that characterizes the key item parts, which can be adopted to supervise the attribute-level representation learning, and hence promote the model's performance as well as interpretability. Thus, in this chapter, we aim to fulfill the fine-grained outfit compatibility modeling by incorporating the semantic attributes of fashion items.

However, fulfilling this goal is nontrivial due to the following challenges. (1) The fashion item attribute labels are not unified or aligned. In other words, each item may have different attribute labels. For instance, as shown in Fig. 5.1, one T-shirt is labeled with attributes of price, sleeve length, design, and brand; while the other has color, material, brand, and price. Thereby, how to fully take advantage of these irregular attribute labels to partially supervise the attribute-level representation learning of fashion items poses a considerable challenge. (2) When disentangling the entire visual embedding into multiple attribute-level representations, how to ensure information intactness during the disentanglement is another challenge. (3) To comprehensively capture the compatibility among fashion items, we incorporate both the coarse-grained item-level and fine-grained attribute-level information into the compatibility modeling. Accordingly, how to seamlessly combine multiple granularities to strengthen the learning performance constitutes another tough challenge.

To address the aforementioned challenges, we present a partially supervised compatibility modeling scheme, called PS-OCM. As shown in Fig. 5.2, it consists of three key components: (1) partially supervised attribute-level embedding learning, (2) disentangled completeness regularization, and (3) hierarchical outfit compatibility modeling. Specifically,

Brand: Under Armour
Sleeve length: Short-Sleeve
Design: V-Neck T-shirt
Price: US $18.90

Brand: Champion
Color: Classic black
Material: Cotton T-shirt
Price: US $15.00

Fig. 5.1 Illustration of two fashion items and their associated irregular attribute labels

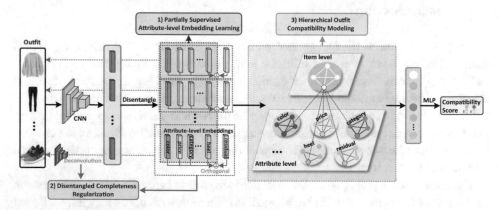

Fig. 5.2 Illustration of our proposed PS-OCM scheme. It consists of three components: partially supervised attribute-level embedding learning, disentangled completeness regularization, and hierarchical outfit compatibility modeling

the first component extracts visual features from each composing item of the given outfit via a pretrained model. It then turns to disentangle the visual feature vector into a set of fine-grained attribute embeddings, which is partially supervised by the irregular attribute labels of each fashion item. The second component works toward an intact disentanglement. This is accomplished by adopting two strategies: orthogonal residual embedding and visual representation reconstruction. An orthogonal residual embedding is introduced to compensate for the information loss, and regularize the orthogonal relationship between the residual embedding and each attribute-level embedding. Additionally, it leverages the deconvolution neural network to ensure that the original image can be reconstructed from the disentangled attribute-level and residual embeddings. The last component contains a hierarchical graph convolutional network, which models the outfit compatibility by jointly integrating the fine-grained attribute-level and coarse-grained item-level information. Ultimately, it fuses the attribute-level compatibility scores and the item-level ones via a multilayer perceptron (MLP) to derive the final compatibility score of the given outfit.

5.2   Related Work

Disentangled representation learning [1] targets learning multiple factorized representations to capture the latent explanatory factors residing in the observed data, which has drawn increasing research attention from various domains, such as the recommendation domain [2, 3] and computer vision domain [4–6]. For example, in the recommendation domain, Hu et al. [7] proposed a novel graph neural news recommendation model with unsupervised preference disentanglement, where a neighborhood routing mechanism is introduced to dynamically identify the latent preference factors affecting the click between a user and a piece of news. In addition, Wang et al. [8] presented a disentangled graph collaborative filtering model to mine the fine-grained user-item relationships. In the computer vision field, Ma et al. [4] strengthened the natural person image generation by disentangling the input image into three intermediate embedding features, corresponding to three main factors: foreground, background, and pose.

As the compatibility relationship among fashion items can be influenced by multiple latent factors, such as color, texture, and style, some researchers also incorporated the disentangled representation to address the task of fashion compatibility modeling. For example, Zheng et al. [9] devised a disentangled graph learning scheme, where the collocation compatibility is disentangled into multiple fine-grained compatibilities among fashion items. Similarly, Guan et al. [10] presented a comprehensive multimodal outfit compatibility modeling scheme, which not only explores the fine-grained outfit compatibility with disentangled item representations but also explicitly models the consistent and complementary correlations between the visual and textual modalities of items. Despite their significant value, the existing efforts mostly overlook the potential of the semantic labels in supervising the disentangled representation learning. Therefore, in this work, we propose utilizing the irregular attributes as partial supervision to guide the disentangled representation learning of items and introducing the completeness regularizer to prevent information loss during disentanglement.

5.3   Methodology

In this section, we first formulate the research problem and then detail the proposed partially supervised outfit compatibility modeling scheme (PS-OCM).

5.3.1 Problem Formulation

In this work, we cast the outfit compatibility modeling task as a binary classification problem, i.e., *whether the given outfit is compatible*. Assume we have a set of N outfits, denoted as $\Omega = \{(O_i, y_i)\}_{i=1}^{N}$, where O_i is the i-th outfit, and y_i denotes its corresponding compatibility label. Specifically, $y_i = 1$ if the outfit O_i is compatible, and $y_i = 0$, otherwise. In addition, we have a set of fashion items $\mathcal{I}$ distributed over T categories. For simplicity, we temporally omit the subscript i of each outfit. An outfit O comprises K fashion items, i.e., $\{I_1, I_2, \ldots, I_K\}$, where $I_i \in \mathcal{I}$ is the i-th composing item of the outfit. Considering that the number of items in an outfit is not fixed, K is a variable. Each item I_i is associated with a visual image V_i and a set of attribute labels $\mathcal{L}_i$. We heuristically predefined a set of attributes (e.g., the *color* and *material*) $\mathcal{A} = \{a_m\}_{m=1}^{M}$ that can be applied to characterize all the fashion items, where a_m is the m-th attribute, and M is the total number of attributes. Moreover, each attribute has a set of corresponding attribute values, e.g., *red* and *blue* are two possible values for the attribute *color*. We then formally use $\mathcal{V}_m = \{v_m^n\}_{n=1}^{N_m}$ to denote all the possible values for the attribute a_m, and N_m is the corresponding total number of values. Therefore, the set of attribute labels of the i-th item can be written as $\mathcal{L}_i = \{l_i^1, l_i^2, \ldots, l_i^M\}$, where $l_i^m \in \mathcal{V}_m$ if the item I_i has m-th attribute; otherwise, $l_i^m = none$. Usually there are two possible reasons leading to $l_i^m = none$: one reason is the intrinsic flaws of the dataset due to loose user-generated annotation, and the other is that items of certain categories essentially cannot present certain attributes (e.g., the *trousers* do not have the attribute of *sleeve length*).

In this work, we target at learning an outfit compatibility model $\mathcal{F}$ to judge whether a given outfit O is compatible. It is formulated as follows,

$$s = \mathcal{F}\big(\{(V_i, \mathcal{L}_i)\}_{i=1}^{K} | \Theta\big), \tag{5.1}$$

where Θ refers to the to-be-learned parameters of our model, and s denotes the compatible probability of the given outfit.

5.3.2 Partially Supervised Compatibility Modeling

As illustrated in Fig. 5.2, PS-OCM consists of three key components: (1) partially supervised attribute-level embedding learning, (2) disentangled completeness regularization, and (3) hierarchical outfit compatibility modeling. We explain them as follows.

Partially Supervised Attribute-Level Embedding Learning
This component aims to derive the fine-grained attribute-level representation of the fashion item, which is the basis for the following hierarchical outfit compatibility modeling. Given an outfit, we first extract the visual feature of each composing item via the convolutional neural networks, which have obtained remarkable success in many computer vision tasks [11, 12]. Specifically, we obtain the overall visual feature embedding of the i-th item in the outfit O as follows,

$$\mathbf{v}_i = \text{CNN}\,(\mathbf{V}_i)\,, \tag{5.2}$$

where $\mathbf{V}_i$ refers to the i-th item image in its raw RGB pixels, $\mathbf{v}_i \in \mathbb{R}^{D_v}$ denotes the extracted visual feature of the i-th item, and D_v is the dimension of the visual feature. In this work, the function CNN refers to ResNet18 [13] pretrained on ImageNet.

As previously mentioned, we predefined a set of M attributes to characterize all the items. Accordingly, we disentangle the visual feature of each item I_i, i.e., $\mathbf{v}_i$, into M attribute-level embeddings. We argue that the attributes are not linearly separable, and hence accomplish this task by the nonlinear MLP mapping. Mathematically, we have

$$\begin{cases} \mathbf{e}_i^1 = \text{MLP}_1\,(\mathbf{v}_i)\,, \\ \mathbf{e}_i^2 = \text{MLP}_2\,(\mathbf{v}_i)\,, \\ \quad\vdots \\ \mathbf{e}_i^M = \text{MLP}_M\,(\mathbf{v}_i)\,, \end{cases} \tag{5.3}$$

where $\mathbf{e}_i^j \in \mathbb{R}^{D_e}$ ($j = 1, \ldots, M$) denotes the j-th disentangled attribute-level embedding of the i-th item, and D_e is the dimension.

Different from existing studies that focus on the unsupervised disentangled representation learning, we argue that even the irregular attribute labels of fashion items contain rich cues. Therefore, they can be used to supervise the attribute-level embedding learning and hence strengthen the final compatibility modeling performance. Thereby, we further utilize M MLPs as the attribute classifiers to explore the attribute labels. As aforementioned, the fashion item attribute labels are irregular. We thus introduce a binary mask $\mathbf{p}_i$ for each item I_i in the outfit to select the available attribute labels of the i-th item. In particular, we define the mask as $\mathbf{p}_i = [p_i^1, p_i^2, \ldots, p_i^M]$, where $p_i^m = \phi(l_i^m)$, and $\phi(\cdot)$ is an indicator function defined as follows,

$$\phi(x) = \begin{cases} 0 & x \text{ is none,} \\ 1 & \text{else.} \end{cases} \tag{5.4}$$

By utilizing the binarized mask, if and only if the item has the corresponding attribute label, we enforce the supervision over the embedding for that attribute. In particular, we adopt the cross-entropy loss to achieve the partial supervision. Formally, for a given outfit O consisting of K items, the partial supervision loss function is formulated as follows,

$$\mathcal{L}_{ps} = \sum_{i=1}^{K} \sum_{m=1}^{M} -\log\left(p\left(l_i^m \mid C^m\left(\mathbf{e}_i^m\right)\right)\right) p_i^m, \tag{5.5}$$

where $C^m(\cdot)$ is the label classifier for the m-th attribute, $\mathbf{e}_i^m$ is the disentangled embedding of the m-th attribute, and l_i^m is the ground-truth attribute label. We illustrate the procedure of partially supervised disentangled attribute-level embedding in Fig. 5.3.

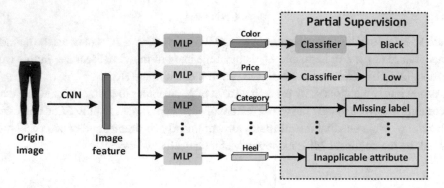

Fig. 5.3 Illustration of the partially supervised attribute-level embedding learning module

Disentangled Completeness Regularization

To prevent information loss during the disentangling process which may degrade the model performance, we devise a disentangled completeness regularizer, as illustrated in Fig. 5.2. In particular, we rely on two strategies to regulate the disentangling process: orthogonal residual embedding and visual representation reconstruction.

Orthogonal Residual Embedding. There may be some implicit visual properties of the item that cannot be represented by the predefined set of attributes. We thus introduce another special attribute *residual* to compensate for the information loss during the disentangled representation learning. Specifically, similar to the M attribute-level embeddings, we adopt another MLP to derive the residual attribute embedding via,

$$\mathbf{e}_i^{M+1} = \mathrm{MLP}_{M+1}(\mathbf{v}_i), \tag{5.6}$$

where $\mathbf{e}_i^{M+1} \in \mathbb{R}^{D_e}$ denotes the residual attribute embedding.

Since the residual attribute embedding acts as compensation for fully representing the item, we argue that it should be complementary to other M attribute-level embeddings that have clear semantics. In other words, the residual embedding should be orthogonal to every other attribute-level embedding. It is worth noting that although we disentangle the visual feature of each fashion item into M attribute-level embeddings, certain embeddings of the given item may be meaningless since some attributes are not universal and cannot be applied to certain items. For example, we can discuss the attribute *sleeve length* for a *T-shirt* but not *trousers*, and the attribute *heel* for a pair of *shoes* rather than a *T-shirt*. Therefore, for each item category, we define a set of meaningful attributes to guarantee effective orthogonal regularization. Thus, we first build the *category-attribute* associations. For the t-th category, we take the union set of attributes used to label items in the t-th category as the whole set of applicable attributes, denoted as $\mathcal{T}_t$. We then introduce a mask $\mathbf{q}_t = [q_t^1, q_t^2, \dots, q_t^M]$ to select the meaningful attributes for the t-th item category, where $q_t^m = 1$ if the predefined m-th attribute belongs to the applicable attribute set $\mathcal{T}_t$; otherwise, $q_t^m = 0$. It is worth noting that in the aforementioned partial supervision module, only the attribute-level embeddings that have corresponding labels are triggered. Whereas in this orthogonal regularization, we further utilize the attribute-level embedding that even has no corresponding label, as long as it can be possibly presented by this item.

Ultimately, we have the following orthogonal regularization,

$$
\begin{aligned}
\mathcal{L}_{or} &= \sum_{i=1}^{K} \sum_{m=1}^{M} \left[\cos \left(\hat{\mathbf{e}}_i^m, \mathbf{e}_i^{M+1} \right) \right]^2 \\
&= \sum_{i=1}^{K} \sum_{m=1}^{M} \left[\cos \left(q_{t_i^*}^m \mathbf{e}_i^m, \mathbf{e}_i^{M+1} \right) \right]^2,
\end{aligned}
\tag{5.7}
$$

where $\cos(\cdot, \cdot)$ is the cosine similarity function, and $t_i^* \in \{1, 2, \ldots, T\}$ refers to the category of the i-th item. It is worth noting that once the m-th attribute cannot be applied to the item I_i, i.e., $q_{t_i^*} = 0$, we ignore the orthogonal regularization between that attribute-level embedding and the residual embedding.

Visual Representation Reconstruction. To avoid information loss during disentangled representation learning, we regulate the disentangled embeddings to be able to reconstruct the original item visual representation. Thus, we feed the concatenation of the meaningful disentangled attribute-level embeddings of the item I_i and the residual one into the deconvolutional neural network [14]. It is formulated as,

$$
\hat{\mathbf{V}}_i = \mathcal{D} \left(\left[q_{t_i^*}^1 \mathbf{e}_i^1 \| q_{t_i^*}^2 \mathbf{e}_i^2 \| \ldots, q_{t_i^*}^M \mathbf{e}_i^M \| \mathbf{e}_i^{M+1} \right] \right),
\tag{5.8}
$$

where the binary masks $\mathbf{q}_{t_i^*}^m$ are used to select the meaningful attribute embeddings of the item I_i, $[\cdot \| \cdot]$ refers to the concatenation operation, $\mathcal{D}(\cdot)$ denotes the deconvolutional neural network, and $\hat{\mathbf{V}}_i$ denotes the reconstructed visual representation of the i-th item. We hereafter utilize l_2 loss to regulate the distance between the reconstructed visual representation and the origin one via,

$$
\mathcal{L}_{rec} = \sum_{i=1}^{K} \left\| \hat{\mathbf{V}}_i - \mathbf{V}_i \right\|_F^2.
\tag{5.9}
$$

Combining the losses of both the orthogonal residual embedding and the visual representation reconstruction constraints, we reach the final loss for regularizing the disentangled completeness as follows,

$$
\mathcal{L}_{dc} = \mathcal{L}_{or} + \mathcal{L}_{rec}.
\tag{5.10}
$$

Hierarchical Outfit Compatibility Modeling

Inspired by previous studies [15, 16], we leverage GCNs to model outfit compatibility. Beyond existing work, we design a novel hierarchical graph convolutional network, which can model the complex compatibility relations among items in an outfit from both attribute and item levels. In particular, the attribute-level compatibility modeling aims to investigate the fine-grained compatibility among fashion items, while the item-level model summarizes the coarse-grained outfit compatibility from the item level.

Attribute-level Compatibility Modeling. Regarding the attribute-level compatibility modeling, given an outfit, we first construct $M + 1$ parallel compatibility modeling graphs $\mathcal{G}_a^m = (\mathcal{N}_a^m, \mathcal{E}_a^m)$, $(m = 1, 2, \ldots, M + 1)$, with each devised to model the outfit compatibility from an attribute aspect[1]. In particular, $\mathcal{N}_a^m$ and $\mathcal{E}_a^m$ refer to the set of nodes and edges, respectively, of the graph $\mathcal{G}_a^m$. In $\mathcal{G}_a^m$ graph, each node refers to a composing item of the outfit that has the corresponding attribute, i.e., a_m. Notably, as previously mentioned, not every attribute can be applied to all the items, e.g., the attribute *sleeve length* cannot be used to characterize a pair of *trousers*. Therefore, for different attributes, different numbers of items are applicable for the attribute-level compatibility modeling. In other words, graphs corresponding to different attributes may have different numbers of nodes. Therefore, for the ease of presentation, we still deploy K item nodes for all these graphs, i.e., $\mathcal{N}_a^m = \{\hat{n}_i^m\}_{i=1}^K$, where $\hat{n}_i^m$ is the i-th node in the $\mathcal{G}_a^m$ graph. However, some nodes in these graphs are defined as the virtual isolated nodes and are inactive during the attribute-level compatibility propagation.

During the learning process, each node $\hat{n}_i^m$ is associated with a hidden state vector $\mathbf{h}_i^m$, which is updated to fulfill the compatibility information propagation over the graph. We initialize the hidden vector of node $\hat{n}_i^m$ by,

$$\mathbf{h}_i^m = \begin{cases} q_{t_i^*}^m \mathbf{e}_i^m, & m \in \{1, 2, \ldots, M\}, \\ \mathbf{e}_i^{M+1}, & m = M + 1. \end{cases} \tag{5.11}$$

Therefore, if the m-th attribute can be applied to the item of the i-th node, we initialize the node with the item's corresponding attribute feature. Otherwise, the node is initialized with an all-zero vector, making it an isolated node in the graph, and it will not join the subsequent compatibility information propagation. Regarding the edge construction for each graph, we introduce an edge between each pair of nonisolated nodes, i.e., each pair of meaningful items in the corresponding attribute-level compatibility modeling.

To simplify the notation, considering that the parallel attribute-level compatibility modeling for different attributes follow the same learning process, we temporally remove all the superscripts m from the above notations and present the general attribute-level compatibility modeling scheme as an example. Inspired by graph attention networks (GAT) [17], we employ the attention mechanism to make each node adaptively absorb compatibility information from the neighbors. Formally, we have

$$\alpha_{ij} = \frac{\exp\left(\mathbf{W}_a \left[\mathbf{h}_i \| \mathbf{h}_j\right]\right)}{\sum_{n_k \in \mathcal{N}_i} \exp\left(\mathbf{W}_a \left[\mathbf{h}_i \| \mathbf{h}_k\right]\right)}, \tag{5.12}$$

where α_{ij} indicates the importance of node n_j's hidden state to node n_i, $\mathbf{W}_a$ is a weight matrix to perform the linear transformation, $[\cdot \| \cdot]$ refers to the concatenation operation, and

---

[1] As previously stated, the residual attribute is also incorporated as a special implicit attribute.

$\mathcal{N}_i$ denotes the neighborhood of node n_i. Once the attention weights α_{ij} are obtained, they are then used to propagate information from the neighbors of node n_i to the node by,

$$\mathbf{h}_i' = \omega \left\{ \mathbf{W}_u \left[\sum_{n_j \in \mathcal{N}_i} \alpha_{ij} \left(\mathbf{h}_i \odot \mathbf{h}_j \right) \right] + \mathbf{b}_u \right\}, \tag{5.13}$$

where $\odot$ denotes the elementwise multiplication, $\mathbf{W}_u$ and $\mathbf{b}_u$ are the parameters of the fully-connected layer, and ω refers to the nonlinear activation function LeakyReLU. The elementwise multiplication $\mathbf{h}_i \odot \mathbf{h}_j$ indicates the compatibility information between the items I_i and I_j. More generally, instead of propagating the features of node n_i's neighbors, we propagate the compatibility information between node n_i and its neighbors, which has proven to be effective in addressing the outfit compatibility modeling task [10].

Based upon the above inference and computation, the updated hidden representation of node n_i is written as,

$$\tilde{\mathbf{h}}_i = \omega \left(\mathbf{W}_o \mathbf{h}_i + \mathbf{b}_o \right) + \mathbf{h}_i', \tag{5.14}$$

where $\mathbf{W}_o$ and $\mathbf{b}_o$ denote the weight matrix and bias to be learned, respectively. The symbol ω denotes the LeakyReLU function. We ultimately feed the updated hidden node embeddings into an MLP to derive the attribute-specific compatibility score of the given outfit via,

$$\begin{cases} c_i = \mathbf{W}_2 \left[\psi \left(\mathbf{W}_1 \tilde{\mathbf{h}}_i + \mathbf{b}_1 \right) \right] + \mathbf{b}_2, \\ c = \frac{1}{K} \sum_{i=1}^{K} c_i, \end{cases} \tag{5.15}$$

where $\mathbf{W}_1, \mathbf{W}_2, \mathbf{b}_1$, and $\mathbf{b}_2$ are the parameters of the MLPs, the symbol ψ denotes the ReLU active function, and c is the compatibility score. Following the above general scheme, we can obtain all the attribute-level compatibility scores, denoted as $\mathbf{c}_u = \left[c^1, c^2, \dots, c^M, c^{M+1} \right]$, as well as the updated hidden attribute-level embeddings of each node/item, i.e., $\tilde{\mathbf{h}}_i = [\tilde{\mathbf{h}}_i^1, \tilde{\mathbf{h}}_i^2, \dots, \tilde{\mathbf{h}}_i^M, \tilde{\mathbf{h}}_i^{M+1}]$.

Item-Level Compatibility Modeling. Similar to attribute-level compatibility modeling, we also construct a compatibility modeling graph $\mathcal{G}_o = (\mathcal{N}_o, \mathcal{E}_o)$ at the overview item level, where $\mathcal{N}_o$ and $\mathcal{E}_o$ refer to the node set and the edge set, respectively. The difference is that we initialize the hidden vector of the i-th node in the graph $\mathcal{G}_o$ from two aspects: the item's original visual feature $\mathbf{v}_i$, and the updated attribute-level item embedding $\tilde{\mathbf{h}}_i^m$'s from the attribute-level compatibility modeling scheme. In this way, a more comprehensive overview representation of the item is derived. Specifically, for the i-th node in the graph $\mathcal{G}_o$, we initialize its hidden vector as follows,

$$\mathbf{g}_i = \left[\mathbf{v}_i \| \mathbf{W}_h \left(\left[\tilde{\mathbf{h}}_i^1 \| \tilde{\mathbf{h}}_i^2 \| \cdots \| \tilde{\mathbf{h}}_i^{M+1} \right] \right) \right], \tag{5.16}$$

where $[\cdot \| \cdot]$ denotes the concatenation operation, and $\mathbf{W}_h \in \mathbb{R}^{D_v \times D_e(M+1)}$ is the to-be-learned weight matrix, which projects the attribute-level embeddings to the same space of the entire visual embeddings. Following the same information propagation scheme as the attribute-level compatibility modeling, we can obtain the item-level compatibility score c_o.

Considering both the attribute- and item-level compatibility modeling results, we feed the concatenation of the attribute- and item-level compatibility scores, i.e., $\mathbf{c} = [c_a \| \mathbf{c}_o]$, into the MLP to obtain the final compatibility probability score as follows,

$$s = \sigma \left\{ \mathbf{W}_4 \left[\psi \left(\mathbf{W}_3 \mathbf{c} + \mathbf{b}_3 \right) \right] + \mathbf{b}_4 \right\}, \tag{5.17}$$

where $\mathbf{W}_3$, $\mathbf{W}_4$, $\mathbf{b}_3$, and $\mathbf{b}_4$ are the parameters of the MLP, the symbol ψ denotes the ReLU active function, and σ refers to the sigmoid active function. We finally adopt the cross-entropy loss to optimize our proposed PS-OCM, and reach the following formulation,

$$\mathcal{L}_{hc} = -ylog(s) - (1 - y)log(1 - s), \tag{5.18}$$

where y is the ground-truth compatibility label for the outfit O. Accordingly, the total loss for our PS-OCM can be written as follows,

$$L = \mathcal{L}_{hc} + \lambda \mathcal{L}_{ps} + \mu \mathcal{L}_{dc}, \tag{5.19}$$

where λ and μ are tradeoff hyperparameters.

Interpretability. The semantic attributes have explicit meaning and can be used naturally to interpret the compatibility evaluation result. In particular, we can identify the prominent attributes that contribute to the final compatibility evaluation most, according to the absolute values of these attribute-specific compatibility scores, i.e., c^ms.

5.4   Experiment

In this section, we first introduce the dataset and experimental settings, and then detail the extensive experiments that we conducted on a real-world dataset by answering the following research questions:

- **RQ1**: Does the proposed PS-OCM outperform the state-of-the-art methods?
- **RQ2**: How does each component affect PS-OCM?
- **RQ3**: What is the intuitive evaluation result of PS-OCM?

5.4.1 Experimental Settings

Dataset and Evaluation Metrics
To justify our model, we resorted to the public dataset IQON3000 [18], due to the fact that each item in IQON3000 has not only the visual image but also several semantic attributes, such as color and category. In particular, IQON3000 consists of 308, 747 outfits, composed

Table 5.1 Attributes, their possible values and total number in the IQON3000 dataset

Attribute	Possible value	Total number
Color	Gray, Back, Green, ⋯	12
Price	Low, Middle, High.	3
Brand	ABISTE, FURLA, BEIGE, ⋯	5, 180
Category	Trousers, Belt, Handbag, ⋯	61
Variety	Coat, Bag, Cosmetics, ⋯	20
Material	Fur, Leather, Denim, ⋯	37
Pattern	Stripe, Embroidery, Animal, ⋯	15
Design	Turtleneck, Frill, Ribbons, ⋯	23
Heel	Chunky, Pin, High, ⋯	6
Dress length	Short, Middle, Long.	3
Sleeve length	Sleeveless, Long, Short, ⋯	4

by 672, 335 items. In total, there are 11 attributes provided by this dataset. Table 5.1 shows the possible value examples and the corresponding number for each attribute. To ensure the dataset quality, we empirically sampled 20, 000 compatible outfits, each of which consisted of at least 2 but no more than 10 items. Since the dataset only provided the compatible outfits, incompatible outfits were composed for training. Specifically, for each compatible outfit, we replaced each of its composing items with a randomly sampled item from the same category to construct the incompatible outfit. In this manner, we end up with a set of 40, 000 compatible/incompatible outfits. We then divided it into the training set, validation set, and test set according to the ratio of 8 : 1 : 1.

Similar to previous studies [10, 15, 16, 19, 20], we justified our proposed PS-OCM scheme with two specific tasks: outfit compatibility estimation and fill-in-the-blank. For these two tasks, we also used the AUC and the accuracy (ACC) as the evaluation metrics, respectively.

Implementation Details

For the image encoder, we employed the ResNet18 [13] pretrained on ImageNet [21] as the backbone, and modified the last layer to make the output feature dimension 256. Pertaining to the MLPs that obtain the disentangled attribute-level embeddings, we set the output dimension to 64. We selected Adam [22] as the training optimizer, with a fixed learning rate of 0.0001. We empirically set the batch size as 32, and both tradeoff hyperparameters, i.e., λ and μ in Eq. (5.19), as 1. All the experiments were implemented by PyTorch over a server equipped with 4 GeForce RTX 2080 Ti GPUs, and the random seeds for model initialization were fixed for the reproducibility.

5.4.2 Model Comparison

To validate the effectiveness of our proposed scheme, we chose the following baselines for comparison, including the pairwise, sequencewise, and graphwise models.

- **Type-aware** [20] devises the type-specific embedding spaces according to the item types, to facilitate the outfit compatibility measurement. The visual-semantic loss is utilized to incorporate the visual and textual information.
- **SCE-NET** [23] embeds the item visual features into multiple semantic subspaces by multiple condition masks, and uses the multimodal features to derive the importance weights for different subspace features to obtain the final item representations.
- **Bi-LSTM** [19] takes the items in an outfit as a sequence ordered by the item categories, and exploits the latent item interaction by a bi-directional LSTM. Notably, the textual information is also adopted to regularize the outfit compatibility modeling by the visual-semantic consistency loss.
- **NGNN** [15] is the first research attempt to employ GNN to model the outfit compatibility, where each outfit is represented as a subgraph, and an attention mechanism is utilized to calculate the outfit compatibility score. For the multimodal features, NGNN designs two graph channels, and the final compatibility score is derived in a late fusion manner.
- **HFGN** [24] develops a hierarchical fashion graph network to jointly fulfill the fashion compatibility modeling and personalized outfit recommendation, where a category-oriented fashion graph is built for each outfit. It only uses the visual features.
- **MM-OCM** [10] explicitly models the consistent and complimentary relations between the visual and textual modalities of fashion items by the parallel and orthogonal regularizations. Moreover, MM-OCM jointly unifies the text-oriented and vision-oriented outfit compatibility modeling with the mutual learning strategy.
- **OCM-CF** [25] directly learns the context-aware global outfit representation by GCNs and the multihead attention mechanism, and employs multiple network branches to explore the hidden complementary factors that affect the outfit compatibility.

Table 5.2 shows the performance of different methods on the outfit compatibility estimation task and fill-in-the-blank task. Notably, the baseline methods were retrained by the released corresponding codes over the IQON3000 dataset. From this table, we make the following observations:

1. The pairwise methods, i.e., Type-aware and SCE-NET, achieved the worst performance on both two tasks. This may be due to the fact that the pairwise methods justify the local compatibility between two items, lacking the global view of the whole outfit.
2. The sequencewise method, i.e., Bi-LSTM, performs better than the pairwise methods, but worse than the graphwise methods, i.e., HFGN and MM-OCM. On the one hand, this confirms the advantage of treating the outfit as a unified sequence rather than the

Table 5.2 Performance comparison between our proposed PS-OCM and other baseline methods on two tasks over the IQON3000 dataset. Notably, the baseline methods were retrained by the released codes. The best results are in bold, while the second best results are underlined

Method	Compat. AUC	FITB ACC
Type-aware [20]	0.6688	0.3901
SCE-NET [23]	0.6792	0.3783
Bi-LSTM [19]	0.7739	0.3813
NGNN [15]	0.7591	0.4002
HFGN [24]	0.8243	0.4511
MM-OCM [10]	0.8444	0.4661
OCM-CF [25]	0.8402	0.4825
PS-OCM	**0.9009**	**0.5412**

item pairs. On the other hand, this implies that treating the outfit as an ordered sequence of fashion items is still suboptimal. This may be attributed to the sequencewise method being able to suffer from the cumulative error propagation problem since it computes the outfit compatibility score by continually predicting the next item with the previous items.

3. Our proposed PS-OCM consistently surpasses all the baseline methods on both tasks. This confirms the advantage of our scheme that utilizes the irregular attribute labels to provide partial supervision to strengthen the item representation learning and employs the hierarchical graph convolutional network to integrate the attribute-level and item-level outfit compatibility learning.

To gain deep insights into our proposed PS-OCM, we further checked the performance of our PS-OCM for outfits with different numbers of composing items on the two tasks. In particular, we reported the performance of our model for outfits with the number of composing items ranging from 2 to 10. As can be seen in Fig. 5.4, our PS-OCM is generally not sensitive to the composing numbers, which indicates that our model PS-OCM can handle the compatibility modeling for outfits with various numbers of items.

5.4.3 Ablation Study

To verify the importance of each component in our model, we conducted ablation experiments on the following derivatives.

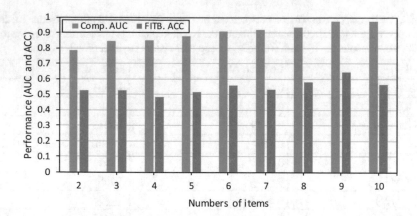

Fig. 5.4 Performance of our PS-OCM on two tasks for outfits with different numbers of items

- **w/o Partial_Supervision**: To explore the effect of the partially supervised attribute embedding learning component, we removed the partial supervision loss by setting $\lambda = 0$ in Eq. (5.19).
- **w/o Orthogonal**: To study the effect of the orthogonal regularization during the visual attributes disentanglement, we removed the orthogonal regularization $\mathcal{L}_{or}$ in Eq. (5.10).
- **w/o Reconstruction**: To validate the necessity of visual representation reconstruction learning, we removed the visual representation reconstruction constraint $\mathcal{L}_{rec}$ in Eq. (5.10).
- **w/o Hierarchical_Graph**: To validate the function of the hierarchical graph compatibility modeling component, we removed this part by directly concatenating the attribute-level embeddings of each outfit to obtain the overall outfit representation and passing it to an MLP to obtain the outfit's compatibility score.
- **Attribute-level_Only**: To verify the importance of coarse-grained item-level information, this derivative only utilizes the fine-grained attribute-level compatibility modeling part in the hierarchical graph compatibility modeling component.
- **Item-level_Only**: Similarly, to justify the necessity of introducing the fine-grained attribute-level compatibility modeling, we removed it from the hierarchical outfit compatibility modeling network.

Based on the ablation experiment illustrated in Table 5.3, we found that our model consistently outperforms all the above derivatives on both tasks, which demonstrates the effectiveness of each component in our proposed PS-OCM. Specifically, we make the following detailed observations.

Table 5.3 Ablation study of our proposed PS-OCM on IQON3000 dataset. The best results are in bold

Method	Compat. AUC	FITB ACC
w/o Partial_Supervision	0.8433	0.4866
w/o Orthogonal	0.8938	0.5293
w/o Deconvolution	0.8909	0.5293
w/o Hierarchical_Graph	0.8197	0.4459
Attribute-level_Only	0.8848	0.5337
Item-level_Only	0.8720	0.5292
PS-OCM	**0.9009**	**0.5412**

1. The performance of w/o Partial_Supervision significantly drops, as compared to PS-OCM, indicating that the partially supervised attribute embedding learning component is indeed helpful to strengthen the visual representation learning performance.
2. Both w/o Orthogonal and w/o Reconstruction are inferior to PS-OCM, which suggests that it is essential to consider the orthogonal regularization and visual feature reconstruction to prevent the visual information loss during the visual feature disentanglement and guarantee the completeness of the disentanglement.
3. w/o Hierarchical_Graph delivers the worst performance, reflecting the overall effectiveness of our proposed hierarchical outfit compatibility modeling component. Moreover, both Attribute-level_Only and Item-level_Only perform better than w/o Hierarchical_Graph, which confirms the necessity of jointly incorporating the attribute-level and item-level compatibility modeling modules. This also reflects that the fine-grained attribute-level features and the overview item-level features complement each other to a certain level toward the outfit compatibility modeling.

As the partial-supervised attribute-level embedding learning contributes the key novelty of our work, we further studied the effect of removing each attribute embedding from the training phase of our PS-OCM. As previously mentioned, we had 12 attributes, including 11 specific attributes in the original dataset and one "residual" attribute we newly defined. Accordingly, we omitted each of the 12 attributes from our model, and hence obtained 12 derivatives of our model, with each named O_{each_attribute}. Figure 5.5 shows the performance of our PS-OCM and its derivatives on the two tasks. As can be seen, removing any specific attribute (e.g., the design or color) hurts our model's performance, which verifies that each specific attribute contributes to the outfit compatibility modeling. In particular, we noticed that the color attribute greatly affects our model's performance on both tasks, which is reasonable, as the color attribute is the most straightforward influential factor in the outfit compatibility modeling. We also found that O_residual underperforms our PS_OCM. This

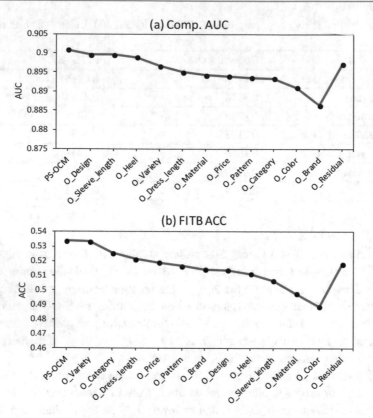

Fig. 5.5 Comparison of the effect of removing a single attribute from our PS-OCM on two tasks

reflects the importance of the residual attribute and indicates its capability of compensating for the information loss during the attribute representation disentanglement.

In addition, we studied the effect of the depth of the GCNs in our hierarchical outfit compatibility modeling component. Figure 5.6 shows the performance of our model with the number of GCN layers ranging from 1 to 5. As can be seen, our model generally performs stably when the number of GCN layers is no more than 3. However, when the number of GCN layers continues to increase, our model's performance decreases. This observation is similar to that reported in [26], and can be attributed to more layers leading to the overfitting problem hurting the model's performance.

5.4.4   Case Study

To obtain an intuitive understanding of our model, we also conducted a case study of our method in the two tasks: outfit compatibility estimation and fill-in-the-blank.

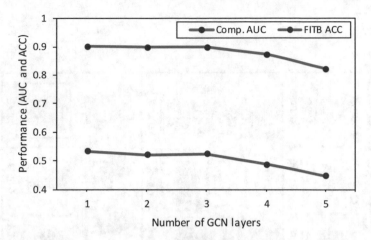

Fig. 5.6 Influence of number of GCN layers on two tasks

Figure 5.7a shows several testing examples of our model on the outfit compatibility esti-mation task, where the importance distribution of attributes, i.e., the normalization of the absolute values of the attribute-level compatibility scores, is also given to intuitively demon-strate the interpretability of our model. As can be seen in the first example, our model yields the correct compatibility estimation and captures the color attribute as the most important influential factor. This is reasonable as the color presented by the outfit is harmonious. In the second example, our model also gives a high compatible probability score and identifies that the pattern attribute is the most important factor. As can be seen, the earrings and the dress in the given outfit do consistently present the dotted pattern. Accordingly, the result makes sense. In the last incompatible example, our PS-OCM gives a low compatibility score, and the pattern attribute is also captured as the most important factor contributing to the incom-patible estimation result. In this example, we found that the striped pattern of the T-shirt, spotted pattern of the skirt, and floral pattern of the sandal form no compatible look.

Figure 5.7b shows several testing examples of our model on the FITB task. In particular, we employed the green tick to refer to the ground-truth item, and the green and red boxes to indicate whether the choice of our model is correct or incorrect, respectively. We also gave the compatibility estimated scores of our model for these candidate items under their corresponding images, respectively. As can be seen from the first two examples in Fig. 5.7b, our model chooses the correct items from the candidate item sets to form compatible outfits with the given query items. Specifically, in the first example, the outfit lacks a top in the sporty style, and our PS-OCM correctly selects the first candidate item by giving it the highest compatibility score with the given query items. In the second example, our model chooses the gray hat, which looks like the most compatible item with the given items as compared

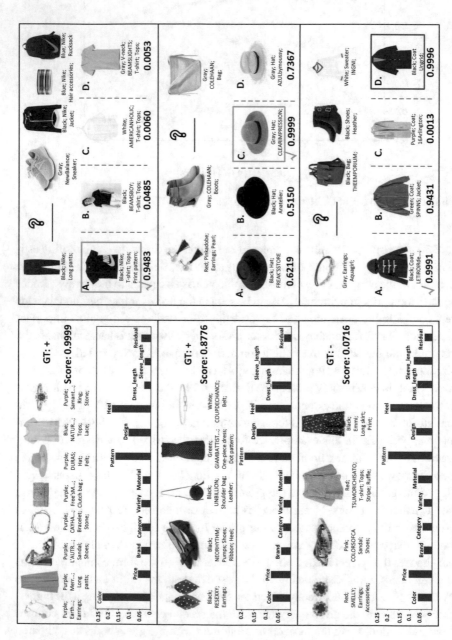

(a) Case study of PS-OCM on the outfit compatibility estimation   (b) Case study of PS-OCM on the fill-in-the-blank task.
task.

Fig. 5.7 Case study of our model on the **a** outfit compatibility estimation task, **b** fill-in-the-blank task

to the other candidate items. Regarding the last example in Fig. 5.7b, although our model fails to give the correct answer by choosing the last candidate item as the most compatible item, we noticed that the chosen item also goes well with the given query items. Overall, these examples show the effectiveness of our model for outfit compatibility modeling.

5.5 Summary

In this chapter, for outfit compatibility modeling, we presented a novel partially supervised compatibility modeling, named PS-OCM, which consists of three key components: (1) partially supervised attribute embedding learning, (2) disentangled completeness regularization, and (3) hierarchical outfit compatibility modeling. In particular, we first presented a partially supervised disentangled learning method to disentangle the visual representation of each item into several attribute-level embeddings, where the irregular attribute labels of fashion items are used as the supervision to strengthen the visual representation learning of items. In addition, we devised the disentangled completeness regularization, including the orthogonal residual embedding and visual representation reconstruction, to prevent information loss during disentanglement. Finally, we designed a hierarchical graph convolutional network that jointly performs attribute- and item-level compatibility modeling. Extensive experiments were conducted on a real-world dataset with two popular tasks: outfit compatibility prediction and fill-in-the-blank. The encouraging experiment results validate the superiority of our proposed model and the importance of each component. In addition, we found that our PS-OCM is not sensitive to the number of items in the outfit, and removing each attribute, including the introduced residual attribute, from the embedding disentanglement will hurt the model's performance. This shows that each attribute affects the outfit compatibility modeling to some extent.

References

1. Jianxin Ma, Peng Cui, Kun Kuang, Xin Wang, and Wenwu Zhu. 2019. Disentangled Graph Convolutional Networks. In *Proceedings of International Conference on Machine Learning*, Vol. 97, 4212–4221. PMLR.
2. Jianxin Ma, Chang Zhou, Peng Cui, Hongxia Yang, and Wenwu Zhu. 2019b. Learning Disentangled Representations for Recommendation. In *Advances in Neural Information Processing Systems*, 5712–5723.
3. Yongfeng Zhang, Guokun Lai, Min Zhang, Yi Zhang, Yiqun Liu, and Shaoping Ma. 2014. Explicit Factor Models for Explainable Recommendation Based on Phrase-Level Sentiment Analysis. In *Proceedings of the International ACM SIGIR Conference on Research and Development in Information Retrieval*, 83–92. ACM.
4. Liqian Ma, Qianru Sun, Stamatios Georgoulis, Luc Van Gool, Bernt Schiele, and Mario Fritz. 2018. Disentangled Person Image Generation. In *IEEE Conference on Computer Vision and Pattern Recognition* 99–108. IEEE.

5. Chen, Hao, Yongjian Deng, Youfu Li, Tzu-Yi. Hung, and Guosheng Lin. 2020. RGBD Salient Object Detection Via Disentangled Cross-Modal Fusion. *IEEE Transactions on Image Processing* 29 (2020): 8407–8416.

6. Yang, Fu.-En., Jing-Cheng. Chang, Chung-Chi. Tsai, and Yu-Chiang Frank. Wang. 2020. A Multi-Domain and Multi-Modal Representation Disentangler for Cross-Domain Image Manipulation and Classification. *IEEE Transactions on Image Processing* 29 (2020): 2795–2807.

7. Linmei Hu, Siyong Xu, Chen Li, Cheng Yang, Chuan Shi, Nan Duan, Xing Xie, and Ming Zhou. 2020. Graph Neural News Recommendation with Unsupervised Preference Disentanglement. In *Proceedings of the Association for Computational Linguistics*, 4255–4264. ACL.

8. Xiang Wang, Hongye Jin, An Zhang, Xiangnan He, Tong Xu, and Tat-Seng Chua. 2020. Disentangled Graph Collaborative Filtering. In *Proceedings of the International ACM SIGIR Conference on Research and Development in Information Retrieval*, 1001–1010. ACM.

9. Na Zheng, Xuemeng Song, Qingying Niu, Xue Dong, Yibing Zhan, and Liqiang Nie. 2021. Collocation and Try-on Network: Whether an Outfit is Compatible. In *Proceedings of the International ACM Conference on Multimedia*, 309–317. ACM.

10. Weili Guan, Haokun Wen, Xuemeng Song, Chung-Hsing Yeh, Xiaojun Chang, and Liqiang Nie. 2021. Multimodal Compatibility Modeling via Exploring the Consistent and Complementary Correlations. In *Proceedings of the International ACM Conference on Multimedia*, 2299–2307. ACM.

11. Sun, Tiancheng, Yulong Wang, Jian Yang, and Hu. Xiaolin. 2017. Convolution Neural Networks With Two Pathways for Image Style Recognition. *IEEE Transactions on Image Processing* 26 (9): 4102–4113.

12. Yupeng, Hu., Meng Liu, Su. Xiaobin, Zan Gao, and Liqiang Nie. 2021. Video Moment Localization via Deep Cross-modal Hashing. *IEEE Transactions on Image Processing* 30 (2021): 4667–4677.

13. Kaiming He, Xiangyu Zhang, Shaoqing Ren, and Jian Sun. 2016. Deep Residual Learning for Image Recognition. In *Proceedings of the IEEE Conference on Computer Vision and Pattern Recognition*, 770–778. IEEE Computer Society.

14. Radford, Alec, Luke Metz, and Soumith Chintala. 2016. Unsupervised Representation Learning with Deep Convolutional Generative Adversarial Networks. In *International Conference on Learning Representations*, 1–15.

15. Zeyu Cui, Zekun Li, Shu Wu, Xiaoyu Zhang, and Liang Wang. 2019. Dressing as a whole: Outfit Compatibility Learning Based on Node-wise Graph Neural Networks. In *Proceedings of the World Wide Web Conference*, 307–317. ACM.

16. Guillem Cucurull, Perouz Taslakian, and David Vázquez. 2019. Context-Aware Visual Compatibility Prediction. In *Proceedings of the IEEE Conference on Computer Vision and Pattern Recognition*, 12617–12626. IEEE.

17. Petar Veličković, Guillem Cucurull, Arantxa Casanova, Adriana Romero, Pietro Liò, and Yoshua Bengio. 2018. Graph Attention Networks. In *Proceedings of the International Conference on Learning Representations*, 45–61. OpenReview.net.

18. Xuemeng Song, Xianjing Han, Yunkai Li, Jingyuan Chen, Xin-Shun Xu, and Liqiang Nie. 2019. GP-BPR: Personalized Compatibility Modeling for Clothing Matching. In *Proceedings of the ACM International Conference on Multimedia*, 320–328. ACM.

19. Xintong Han, Zuxuan Wu, Yu-Gang Jiang, and Larry S. Davis. 2017. Learning Fashion Compatibility with Bidirectional LSTMs. In *Proceedings of the ACM International Conference on Multimedia*, 1078–1086. ACM.

20. Mariya I. Vasileva, Bryan A. Plummer, Krishna Dusad, Shreya Rajpal, Ranjitha Kumar, and David A. Forsyth. 2018. Learning Type-Aware Embeddings for Fashion Compatibility. In *European Conference on Computer Vision*, 405–421. Springer.

21. Jia Deng, Wei Dong, Richard Socher, Li-Jia Li, Kai Li, and Fei-Fei Li. 2009. ImageNet: A Large-Scale Hierarchical Image Database. In *Proceedings of the IEEE Conference on Computer Vision and Pattern Recognition*, 248–255. IEEE.
22. Diederik P. Kingma and Jimmy Ba. 2015. Adam: A Method for Stochastic Optimization. In *Proceedings of the International Conference on Learning Representations*, 1–15. OpenReview.net.
23. Reuben Tan, Mariya I. Vasileva, Kate Saenko, and Bryan A. Plummer. 2019. Learning Similarity Conditions Without Explicit Supervision. In *Proceedings of the IEEE International Conference on Computer Vision*, 10372–10381. IEEE.
24. Xingchen Li, Xiang Wang, Xiangnan He, Long Chen, Jun Xiao, and Tat-Seng Chua. 2020. Hierarchical Fashion Graph Network for Personalized Outfit Recommendation. In *Proceedings of the International ACM SIGIR Conference on Research and Development in Information Retrieval*, 159–168. ACM.
25. Tianyu Su, Xuemeng Song, Na Zheng, Weili Guan, Yan Li, and Liqiang Nie. 2021. Complementary Factorization Towards Outfit Compatibility Modeling. In *Proceedings of the ACM International Conference on Multimedia*, 4073–4081. ACM.
26. Xiaolin Chen, Xuemeng Song, Ruiyang Ren, Lei Zhu, Zhiyong Cheng, and Liqiang Nie. 2020. Fine-Grained Privacy Detection with Graph-Regularized Hierarchical Attentive Representation Learning. *ACM Transactions on Information Systems* 38 (4): 37:1–37:26.

Heterogeneous Graph Learning for Personalized OCM

6.1 Introduction

One common limitation of our previous works presented in Chaps. 2, 3, 4, and 5 is that they all evaluate the outfit compatibility from the general standard. In fact, there may be some subjective factors influencing the outfit compatibility evaluation, namely, for the same garment, different users may have different evaluations. In other words, different people usually have different preferences to make their personal ideal outfits, which may be caused by their diverse growing circumstances or educational backgrounds. For example, as shown in Fig. 6.1, given the same pink shirt, user A prefers to match it with a homochromatic skirt and high-heeled shoes; whereas user B likes to coordinate it with casual jeans and white sneakers. Therefore, personalized outfit compatibility modeling, called POCM, considering users' preferences when measuring the compatibility among fashion items, merit our special attention. A few pioneer researchers have noticed this phenomenon and dedicated their efforts to POCM [1–3]. These efforts study the user and item entities, as well as their relations. They, however, overlook another important entity type in POCM, namely, attributes. Conveying rich semantics, attributes play a pivotal role in characterizing items and delivering users' preferences to items. For instance, we may express "I would like to buy a black coat with a fur collar", whereby the key information is conveyed via the semantic attributes. To alleviate such a problem, we incorporate attributes associated with fashion items and work toward fully exploring all the related entities (i.e., users, items, and attributes) and their various relations (i.e., user-item interactions, item-item matching relations, and item-attribute association relations) to promote the POCM performance. Without loss of generality, we specifically study the research problem of "which bottom (top) is compatible to the given top (bottom) for a specific user".

Addressing the aforementioned research is, however, nontrivial due to the following challenges. **C1**: POCM involves three kinds of entities with heterogeneous contents: users,

W. Guan et al., *Graph Learning for Fashion Compatibility Modeling*,
Synthesis Lectures on Information Concepts, Retrieval, and Services,
https://doi.org/10.1007/978-3-031-18817-6_6

Fig. 6.1 Examples of users' outfit compositions shared on the online fashion-oriented website

items, and attributes. Specifically, users are pure IDs, items are composed of images and textual descriptions, while attributes are in the form of textual phrases. Thereby, how to effectively organize these heterogeneous data seamlessly poses the first research challenge. **C2**: Different from the item and attribute entities, we do not have the specific content information of user entities. The conventional user embedding paradigm usually assigns a fixed one-hot embedding or learnable embedding to represent each user. This is actually not applicable to new users arriving during the testing phase, even for the case in which we have the historical interactions of these new users. Accordingly, how to derive the user embedding is another challenge. **C3**: In fact, apart from the direct relations, like the user-item interaction relation, item-item matching relation, and item-attribute association relation, there are also high-order relations among the three types of entities. For example, similar bottoms matching the same top may share some common attributes. Another example is that users with similar tastes tend to like items with similar attributes. Therefore, how to explore the high-order relations among these entities to strengthen the model's performance constitutes the third challenge.

To address the challenge **C1**, we organize the users, items, and attributes in the context of POCM into a unified heterogeneous graph. Specifically, these three kinds of entities are nodes of this graph. The nodes are linked by three kinds of edges, which are user-item interactions, item-item matching relations, and item-attribute association relations. It is worth mentioning that in this graph there is no direct edge linking the user and attribute entities. We then devise a novel metapath-guided personalized compatibility modeling scheme to address **C2** and **C3**, named as MG-POCM, as shown in Fig. 6.2. This scheme consists of three key

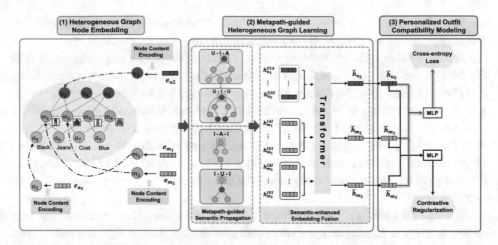

Fig. 6.2 Illustration of the proposed MG-POCM scheme. It consists of three key components: (1) heterogeneous graph node embedding, (2) metapath-guided heterogeneous graph learning, and (3) personalized outfit compatibility modeling

components: heterogeneous graph node embedding, metapath-guided heterogeneous graph learning, and personalized outfit compatibility modeling. The first component works on embedding each type of entity in our heterogeneous graph. To represent users, we devise a multimodal content-oriented user embedding module, which derives the user embedding based on the multimodal contents of his/her interacted items, a straightforward cue indicating the user's preference. As to the second component, we first define multiple user-oriented and item-oriented metapaths (e.g., User → Item → User and Item → Attribute → Item) to capture the high-order relations among entities, which naturally resolves the third challenge **C3**. Thereafter, we conduct the multiple metapath-guided heterogeneous graph learning to obtain the multiple semantic-enhanced user/item embedding of each user/item, whereby each metapath corresponds to a specific semantic. A transformer [4] is used to adaptively fuse the semantic-enhanced user/item embeddings under different metapaths for each user/item. Ultimately, in the last component, in addition to the typical cross-entropy loss, we also introduce the contrastive regularization to enhance embedding learning.

6.2   Related Work

This work is related to personalized compatibility modeling and heterogeneous graph learning.

Personalized Compatibility Modeling. Despite the significant progress made by previous outfit compatibility modeling efforts, they purely focus on the general item-item compatibility and overlook users' preferences in the fashion compatibility estimation. In fact, for the same fashion outfit, different users may have different evaluation results. Inspired by this, some studies have resorted to personalized outfit compatibility modeling. For example, a personalized compatibility modeling scheme for personalized clothing matching, named GP-BPR, is presented in [1], which jointly considers the general (item-item) compatibility and personal (user-item) preference for personalized clothing matching. Both the image and context description of items are utilized in the comprehensive modeling. Moving a step forward, Sagar et al. [3] introduced an attributewise interpretable personal preference modeling scheme to strengthen the model interpretability whereby the images and textual descriptions of items are explored. Additionally, Li et al. [2] developed a hierarchical fashion graph network to simultaneously model the rich relationships among users, items, and outfits.

Although these efforts have achieved compelling success, they overlook the item attributes when estimating compatibility. Attributes express the key item semantics items and reflect the specific user preferences. As a complementary effort, in this book, we incorporate the attribute entities and their semantic contents to comprehensively study the POCM problem.

Heterogeneous Graph Learning. Due to the ubiquity of heterogeneous graphs in the real-world setting, containing multiple types of nodes and relations among these nodes [5, 6], increasing research efforts have been dedicated to heterogeneous graph learning. In a sense, existing methods focus on the heterogeneous graph embedding via learning a powerful low-dimensional vector representation for each node to benefit the potential downstream applications, such as node classification [7, 8] and personalized recommendation [9, 10]. To accomplish this task, previous methods mostly rely on the metapath [11], i.e., a sequence of node and edge types, delivering certain semantic information of the graph. For example, Dong et al. [12] developed metapath-based random walks to construct the heterogeneous neighborhood of a node and then utilized a skip-gram model [13] to perform node embeddings. One key limitation of this method is that it only utilizes a single metapath, which may be insufficient to cover all useful information. To address this issue, Shi et al. [14] designed a novel strategy to generate the meaningful node sequences and utilized fusion functions to learn node representation. In addition, Zhang et al. [15] introduced a heterogeneous graph neural network model, named HetGNN, to jointly explore the heterogeneous structures and contents of each node. To obtain superior node representation, several researchers [16–18] have utilized the attention mechanism to select the most useful metapath. For example, Wang et al. [17] proposed a heterogeneous graph attention network, which incorporates both node- and semantic-level attention to learn the importance of nodes and metapaths in the node embedding. Subsequently, Zhang et al. [18] proposed an attentive heterogeneous graph neural network for heterogeneous graph embedding, where the node-level attention is considered, and a semantic-level neural network is utilized rather than the semantic-level attention for capturing the feature interaction among node embeddings under different metapaths. Differently, Xing et al. [19] regarded each metapath as a specific view, and borrowed the idea of multiview learning to comprehensively encode the node representations of different views into a latent representation. To address the practical issue of missing attributes, Jin et al. [20] proposed a general framework for heterogeneous graph neural network via attribute completion, comprising two key components: prelearning topological embedding and attribute completion with attention mechanism.

Inspired by the great success of these methods on heterogeneous graph learning, we seamlessly organize the various entities and relations in the context of POCM into a unified heterogeneous graph. It is worth emphasizing that we design task-specific metapaths and creatively incorporate the transformer to fuse the semantic-enhanced user/item embeddings.

6.3 Methodology

In this section, we first formulate the research problem and then detail the three components of our proposed MG-POCM scheme.

6.3.1 Problem Formulation

Formally, we first clarify the notations. We use bold uppercase letters (e.g., $\mathbf{W}$) and bold lowercase letters (e.g., $\mathbf{b}$) to represent matrices and vectors, respectively. All vectors are in column forms. Additionally, we employ nonbold letters (e.g., W and W) to denote scalars and Greek letters (e.g., α) to represent regularization parameters.

In this work, we focus on fulfilling the task of POCM. Without loss of generality, we study the particular problem of "whether the given bottom (top) matches the given top (bottom) and together compose a favorable outfit for the given user". Assume that we have a set of N_u users $\mathcal{U} = \{u_1, u_2, \ldots, u_{N_u}\}$, and a set of N_m items $\mathcal{M} = \{m_1, m_2, \ldots, m_{N_m}\}$. For an arbitrary item $m_i, i = 1, 2, \ldots, N_m$, it is composed of an image v_i, a textual description t_i, and a set of attributes $\mathcal{A}_i \subseteq \mathcal{A}$, where $\mathcal{A} = \bigcup_{i=1}^{N_m} \mathcal{A}_i = \{a_1, a_2, \ldots, a_{N_a}\}$ represents the entire attribute set in the form of semantic phrases, like *red color*, *wool material*, and *V-neck design*. The symbol N_a denotes the total number of attributes in our dataset. To simplify the formulation, in this work, we only consider the tops and bottoms. Therefore, the set of items can be rewritten as $\mathcal{M} = \mathcal{M}^t \cup \mathcal{M}^b$, where $\mathcal{M}^t$ and $\mathcal{M}^b$ refer to the sets of tops and bottoms, respectively. Each user u is historically associated with a set of top-bottom pairs $\mathcal{X}^u = \{(m_{t_1}^u, m_{b_1}^u), (m_{t_2}^u, m_{b_2}^u), \ldots, (m_{t_{M_u}}^u, m_{b_{M_u}}^u)\}$, where $m_{t_*}^u \in \mathcal{M}^t, m_{b_*}^u \in \mathcal{M}^b$, and M_u denotes the total number of interacted top-bottom pairs by the user u. We resort to a heterogeneous graph to organize the complicated entities and relations within a unified structure. In particular, we denote the graph as $\mathcal{G} = (\mathcal{E}, \mathcal{R})$, where $\mathcal{E} = \mathcal{U} \cup \mathcal{M} \cup \mathcal{A}$ denotes the set of entity nodes, consisting of user entities, item entities, and attribute entities, while $\mathcal{R}$ denotes the set of edges linking nodes to characterize various relations among entities, i.e., user-item historical interactions, item-attribute association relations, and item-item matching relations. Ultimately, we work in learning the following compatibility estimation function,

$$p_{ij}^k = \mathcal{F}(m_k \in \mathcal{M}^{b(t)} | u_i, m_j \in \mathcal{M}^{t(b)}), \tag{6.1}$$

where p_{ij}^k denotes the compatibility degree of a bottom (top) m_k to the given top (bottom) m_j for the user u_i.

6.3.2   Metapath-Guided Personalized Compatibility Modeling

As illustrated in Fig. 6.2, MG-POCM consists of three components: (1) heterogeneous graph node embedding, (2) metapath-guided heterogeneous graph learning, and (3) personalized outfit compatibility modeling. In this subsection, we explain each of them.

Heterogeneous Graph Node Embedding.
This component aims to derive the initial node-level representations in the heterogeneous graph. The heterogeneous graph has three types of entities and the node contents differ remarkably. Therefore, we learn their embeddings separately as shown in Fig. 6.3.

   Item Entity Embedding. Each item entity is composed of an image and a textual description. The multimodal cues of each item mutually complement each other. For the arbitrary item m_i, regardless of its category (i.e., top or bottom), we utilize the ResNet, which has shown compelling success in many computer vision tasks [21], to extract its visual feature. We adopt the pretrained BERT to obtain its textual feature[1] due to its prominent performance in textual representation learning [22]. Specifically, we employ the averaged hidden states corresponding to the special token attached at the beginning of the input sequence, i.e., [CLS], of the last two layers of BERT as the textual description representation. Finally, we concatenate the visual and textual features of each item to derive the final item embedding, and use a learnable fully-connected layer to project the item embedding into a lower dimensional space. Mathematically, we have

$$\begin{cases} \mathbf{e}_{v_i} = \text{ResNet}\,(v_i)\,, \\ \mathbf{e}_{t_i} = \text{BERT}\,(t_i)_{[CLS]}\,, \\ \mathbf{e}_{m_i} = f_t\left([\mathbf{e}_{v_i}, \mathbf{e}_{t_i}]\right), \end{cases} \tag{6.2}$$

where $\mathbf{e}_{v_i} \in \mathbb{R}^{D_v}$ and $\mathbf{e}_{t_i} \in \mathbb{R}^{D_t}$ refer to the visual and textual embedding of the item m_i, respectively. Accordingly, the symbols D_v and D_t are the dimensions of the visual and textual embeddings, respectively. ResNet and BERT denote the corresponding neural networks. [,] refers to the concatenation operation, f_t denotes the learnable fully-connected layer, and $\mathbf{e}_{m_i} \in \mathbb{R}^D$ is the final embedding of the item m_i.

   Attribute Entity Embedding. To fully utilize the semantic content of each attribute entity, we also resort to the pretrained BERT with a learnable fully-connected layer to derive its embedding instead of using the one-hot vector or treating it as the learnable parameter.

---

[1] Before feeding a text into the BERT, the text is first tokenized into standard vocabularies.

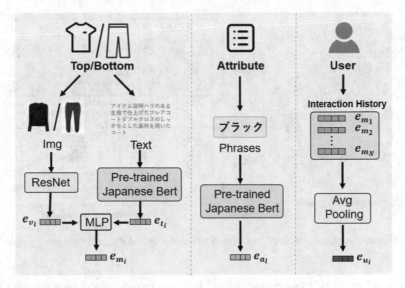

Fig. 6.3 Illustration of the heterogeneous graph node embedding component

Notably, for attribute embedding, we only adopt the representation of the special token [CLS] from the last layer of BERT due to its shorter length, as compared with the textual description. Formally, for each attribute entity a_l, we obtain its embedding as follows,

$$\mathbf{e}_{a_l} = f_a(\text{BERT}(a_l)_{[CLS]}), \qquad (6.3)$$

where $\mathbf{e}_{a_l} \in \mathbb{R}^D$ denotes the initial embedding of the attribute entity a_l, and f_a denotes the fully-connected layer for embedding fine-tuning.

User Entity Embedding. Instead of using the one-hot embeddings, we resort to aggregating all the embeddings of the user's one-hop neighbor nodes (i.e., all the items interacted by the user before) to derive the initial embedding of each user entity. The underlying philosophy is two-fold: (1) the items that are historically interacted by users signal users' preferences and tastes, and (2) the embedding of a cold-start user can also be derived as long as his/her interacted items appeared before. Specifically, we determine the user embedding below,

$$\mathbf{e}_{u_i} = \frac{1}{|\mathcal{N}^{u_i}|} \sum_{m_i \in \mathcal{N}^{u_i}} \mathbf{e}_{m_i}, \qquad (6.4)$$

where $\mathbf{e}_{u_i} \in \mathbb{R}^D$ denotes the user embedding u_i, and $\mathcal{N}^{u_i}$ refers to the set of one-hop neighbors of the user entity u_i.

Metapath-guided Heterogeneous Graph Learning

In this component, we conduct the metapath-guided heterogeneous graph representation learning to refine each entity's embedding with their context information. In particular, we

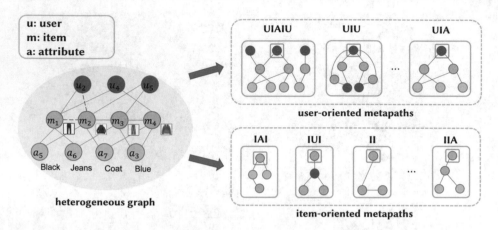

Fig. 6.4 Illustration of user-oriented and item-oriented metapaths via heterogeneous graph

first define a few user-/item-oriented metapaths to capture the high-order relations among entities, and then perform the metapath-guided semantic propagation to derive multiple semantic-enhanced embeddings for each user/item entity. Therein, each applicable metapath corresponds to a specific semantic-enhanced embedding. Ultimately, we fuse all the semantic-enhanced embeddings via a transformer to obtain the final user/item representation.

User-/Item-oriented Metapath Definition. According to [11], a metapath is defined as a path in the form of $X_1 \xrightarrow{R_1} X_2 \xrightarrow{R_2} \cdots \xrightarrow{R_n} X_{n+1}$, which describes a composite relation between entities. In our work, as illustrated in Fig. 6.4, there are actually various metapaths residing in our constructed heterogeneous graph, whereby three entities and rich relations exist. Intuitively, different metapaths reflect different semantics. For example, the metapath UIA[2] implies that a user historically prefers an item and that item possesses an attribute, while UIU indicates that these two end users like the same fashion item. Analogously, the metapath IAI refers to that the two end items share the same attribute, while IUI conveys that the two end items are interacted with by the same user. Pertaining to the POCM context, we only adopt metapaths that start from user entities and item entities. Formally, let $\mathcal{P}_{user} = \{r_1, \ldots, r_Y\}$ and $\mathcal{P}_{item} = \{s_1, \ldots, s_Z\}$ denote the set of predefined user-oriented and item-oriented metapaths, respectively. Y and Z denote the total number of user-oriented and item-oriented metapaths, respectively.

Metapath-guided Semantic Propagation. Based on the predefined user- and item-oriented metapaths, we can derive the corresponding metapath-guided user-oriented subgraphs for each user entity and the item-oriented subgraphs for each item entity via the breadth-first search strategy. Thereafter, based on the different information encoded by different metapaths, we can learn the user/item entity's embeddings with different semantics.

---

[2] Due to the limited space, we omit the relation types between entities.

To intuitively clarify how to refine users' or items' embeddings, we take the metapath UIA as an example. Other metapath-guided learning repeats the same procedure.

Assume that the metapath UIA is applicable to the user entity u_i. We then build a subgraph $\mathcal{G}_{u_i}^{\text{UIA}}$ for the user entity u_i. Since the length of the metapath UIA is three, we denote the one-hop neighbors of user entity u_i as $\mathcal{N}_{u_i}^{\text{UIA}(1)}$ consisting of all the items the user once interacted with. In the same way, we denote the two-hop neighbors of the user entity u_i as $\mathcal{N}_{u_i}^{\text{UIA}(2)}$, comprising all the attributes associated with items in $\mathcal{N}_{u_i}^{\text{UIA}(1)}$. Following that, we first aggregate the information from the two-hop neighbors to enhance the one-hop neighbors' embeddings, and then based on that learn the user's semantic-enhanced embedding as follows:

$$
\begin{cases}
\mathbf{e}_{m_i}^{\text{UIA}} = \mathcal{H}\left(\mathbf{e}_{a_l} | a_l \in \mathcal{N}_{u_i}^{\text{UIA}(2)}\right), m_i \in \mathcal{N}_{u_i}^{\text{UIA}(1)}, \\
\mathbf{h}_{u_i}^{\text{UIA}} = \mathcal{H}\left(\mathbf{e}_{m_i}^{\text{UIA}} | m_i \in \mathcal{N}_{u_i}^{\text{UIA}(1)}\right),
\end{cases}
\tag{6.5}
$$

where $\mathcal{H}$ is the aggregation function, and $\mathbf{e}_{m_i}^{\text{UIA}}$ denotes the semantic-enhanced embedding of the item entity m_i, which is an one-hop neighbor of the user entity u_i. $\mathbf{h}_{u_i}^{\text{UIA}}$ represents the semantic-enhanced embedding of the user entity u_i. It is worth noting that during each hop aggregation, different neighbors may play different roles in characterizing the center entity. Specifically, some attributes may be more important in conveying the item's properties, while some items may contribute more in reflecting the user's preference. Therefore, we adopt the graph attention mechanism of GAT [23] as the aggregation function, to highlight the informative and meaningful neighbor nodes. For simplicity, we take the aggregation operation over the one-hop neighbors of the user entity u_i as an example, and that over the two-hop neighbors can be defined similarly. Specifically, the aggregation operation $\mathcal{H}$ over the one-hop neighbors of the user entity u_i can be written as follows,

$$
\begin{cases}
\mathbf{h}_{u_i}^{\text{UIA}} = \mathbf{e}_{u_i} + \sigma \left(\sum_{m_j \in \mathcal{N}_{u_i}^{\text{UIA}}} \alpha_{ij} \mathbf{e}_{m_j} \right), \\
\alpha_{ij} = \dfrac{\exp\left(\sigma\left(\mathbf{W}^{\text{UIA}}[e_{u_i}, e_{m_j}]\right)\right)}{\sum_{m_j \in \mathcal{N}_{u_i}^{\text{UIA}}} \exp\left(\sigma\left(\mathbf{W}^{\text{UIA}}[e_{u_i}, e_{m_j}]\right)\right)},
\end{cases}
\tag{6.6}
$$

where $\sigma(\cdot)$ denotes the activation function, $[,]$ refers to the concatenation operation, and $\mathbf{W}^{\text{UIA}} \in \mathbb{R}^{2D*1}$ is the node-level attention vector for the information aggregation under the metapath UIA.

Theoretically, repeating the above process for semantic propagation for all the other user-oriented metapaths, we can derive Y semantic-enhanced user embeddings for user entity u_i. However, in practice, not every user-oriented (item-oriented) metapath can be applied to a given user (item) entity. For example, once a user shares no preferred item with other users, we cannot derive the subgraph according to the metapath UIU. Accordingly, we use $\mathcal{P}^{u_i} = \{r_{i_1}, \ldots, r_{i_{Y_i}}\}$ to denote the set of metapaths that can be applied to the user entity u_i, where Y_i refers to the total number of metapaths applicable to the user u_i, and $r_{i_n} \in \mathcal{P}_{user}, n = 1, \ldots, Y_i$. Based upon $\mathcal{P}^{u_i}$, we can derive the corresponding semantic-enhanced embeddings for the user entity u_i, termed as $\{\mathbf{h}_{u_i}^p | p \in \mathcal{P}^{u_i}\}$, following the above metapath-guided semantic propagation process. Similarly, we use $\mathcal{P}^{m_i} = \{s_{i_1}, \ldots, s_{i_{Z_i}}\}$ to denote the set of metapaths that can be applied to the item entity m_i, where Z_i is the total number of metapaths applicable to the item m_i, and $s_{i_z} \in \mathcal{P}_{item}, z = 1, \ldots, Z_i$. In the same manner, we reach the semantic-enhanced embeddings for the item entity m_i as $\{\mathbf{h}_{m_i}^p | p \in \mathcal{P}^{m_i}\}$.

Semantic-enhanced Embedding Fusion. Thus far, we have achieved multiple semantic-enhanced embeddings for each user and item entity under different metapaths, and each embedding characterizes one aspect. To comprehensively represent each user or item, we propose fusing the multiple embeddings of each user or item.

In particular, we leverage the transformer [4] without the positional coding to perform the multisemantic embedding fusion due to the following two concerns: (1) the number of semantic-enhanced embeddings for different users can be different, and (2) there is no explicit order among these semantic-enhanced embeddings of each user or item entity. To ensure that the fused embeddings of the users and items are in the same space, we adopt a single transformer to fulfill both user and item entities' embedding fusion as follows,

$$\begin{cases} \tilde{\mathbf{h}}_{u_i} = \text{Transformer}(\mathbf{h}_{u_i}^p | p \in \mathcal{P}^{u_i}) \\ \tilde{\mathbf{h}}_{m_i} = \text{Transformer}(\mathbf{h}_{m_i}^p | p \in \mathcal{P}^{m_i}) \end{cases} \tag{6.7}$$

where $\tilde{\mathbf{h}}_{u_i}$ and $\tilde{\mathbf{h}}_{m_i}$ are the final representation for the user u_i and item m_i, respectively.

Personalized Outfit Compatibility Modeling

To accomplish the POCM task, we first build the training set $\Omega = \{(u_i, m_j, m_{k+}, m_{k-}) | m_j \in \mathcal{M}^{t(b)}, m_{q+}, m_{q-} \in \mathcal{M}^{b(t)}, y_{ij}^{k+} = 1, y_{ij}^{k-} = 0\}$, where $y_{ij}^{k+} = 1$ denotes the triplet (u_i, m_j, m_{k+}) is compatible, i.e., the item m_{k+} goes well with the given item m_j according to user u_i's preference. $y_{ij}^{k-} = 0$ indicates that the triplet (u_i, m_j, m_{k-}) is incompatible. Following that, for each triplet, we obtain each entity's representation according to Eq. (6.7), namely, $\tilde{\mathbf{h}}_{u_i}$, $\tilde{\mathbf{h}}_{m_j}$, and $\tilde{\mathbf{h}}_{m_{k+}} / \tilde{\mathbf{h}}_{m_{k-}}$. We then resort to the MLP to derive the compatibility score for each triplet as follows,

$$\hat{p}_{ij}^{k+(-)} = \text{MLP}_0([\tilde{\mathbf{h}}_{u_i}, \tilde{\mathbf{h}}_{m_j}, \tilde{\mathbf{h}}_{m_{k+(-)}}]), \tag{6.8}$$

where $\hat{p}_{ij}^{k+(-)}$ is the predicted compatibility score for the given triplet. We then adopt the cross-entropy loss as follows,

$$\mathcal{L}^{(i,j,k+,k-)} = -log\left(\frac{exp(\hat{p}_{ij}^{k+})}{exp(\hat{p}_{ij}^{k+}) + exp(\hat{p}_{ij}^{k-})}\right). \tag{6.9}$$

Intuitively, the compatible and incompatible triplets should follow some compatible and incompatible patterns, respectively. In light of this, given a compatible triplet (u_{i+}, m_{j+}, m_{k+}), we argue that its latent representation should be more similar to that of a compatible triplet as compared to that of an incompatible one (u_{i-}, m_{j-}, m_{k-}). Accordingly, we further introduce contrastive regularization to regulate the similarity between latent representations of different triplet pairs. Assume that $p_1^+ = (u_{i_1^+}, m_{j_1^+}, m_{k_1^+})$ and $p_2^+ = (u_{i_2^+}, m_{j_2^+}, m_{k_2^+})$ are two compatible triplets, while $n^- = (u_{i-}, m_{j-}, m_{k-})$ is an incompatible triplet. We utilize two MLPs to obtain the latent representations for these three triplets as follows,

$$
\begin{cases}
\tilde{\mathbf{h}}_{p_1^+} = \text{MLP}_1([\tilde{\mathbf{h}}_{u_{i_1^+}}, \tilde{\mathbf{h}}_{m_{p_1^+}}, \tilde{\mathbf{h}}_{m_{q_1^+}}]), \\
\tilde{\mathbf{h}}_{p_2^+} = \text{MLP}_2([\tilde{\mathbf{h}}_{u_{i_2^+}}, \tilde{\mathbf{h}}_{m_{p_2^+}}, \tilde{\mathbf{h}}_{m_{q_2^+}}]), \\
\tilde{\mathbf{h}}_{n^-} = \text{MLP}_2([\tilde{\mathbf{h}}_{u_i^-}, \tilde{\mathbf{h}}_{m_{p}^-}, \tilde{\mathbf{h}}_{m_{q}^-}]),
\end{cases}
\tag{6.10}
$$

where $\tilde{\mathbf{h}}_{p_1^+}$ and $\tilde{\mathbf{h}}_{p_2^+}$ are the latent representations of the two compatible/positive triplets, while $\tilde{\mathbf{h}}_{n^-}$ is the latent representation of the incompatible/negative triplet. We then use the following contrastive regularization as follows,

$$
\mathcal{L}_{cons}^{(p_1^+, p_2^+, n^-)} = -log \frac{exp(sim(\tilde{\mathbf{h}}_{p_1^+}, \tilde{\mathbf{h}}_{p_2^+}))}{exp(sim(\tilde{\mathbf{h}}_{p_1^+}, \tilde{\mathbf{h}}_{p_2^+})) + exp(sim(\tilde{\mathbf{h}}_{p_1^+}, \tilde{\mathbf{h}}_{n^-}))},
\tag{6.11}
$$

where $sim(,)$ refers to the dot product operation. Finally, our objective function can be written as follows,

$$
\mathcal{L} = \sum_{(u_i, m_j, m_{k+}, m_{k-})} \mathcal{L}^{(i, j, k+, k-)} + \lambda \sum_{(p_1^+, p_2^+, n^-)} \mathcal{L}_{cons}^{(p_1^+, p_2^+, n^-)},
\tag{6.12}
$$

where λ is the nonnegative hyperparameter balancing the importance of the cross-entropy loss and contrastive regularization.

6.4 Experiment

In this section, we conducted experiments over real-world datasets by answering the following research questions.

- **RQ1**: Does MG-POCM outperform state-of-the-art baselines?
- **RQ2**: How does each module affect MG-POCM?
- **RQ3**: Is our model sensitive to the number of the transformer and GAT layers?
- **RQ4**: What is the intuitive performance of MG-POCM?

6.4.1 Experimental Settings

In this part, we present the dataset, evaluation tasks, metrics, and the implementation details.

Dataset.

To justify our model, similar to existing methods [1, 3], we also resorted to the public benchmark dataset IQON3000 [1], due to the fact that each item in IQON3000 has not only the visual image and textual description but also the semantic attributes, such as the color and category. In particular, IQON3000 consists of 308, 747 outfits, composed by 672, 335 items. To fit our task and ensure the quality of the dataset, we did not completely follow up the experimental setting in [1, 3] considering the following two concerns. (1) As to a given user, they only focus on matching bottoms for a given top. By contrast, in our work, the top and bottom are arbitrarily switchable for a given user. That is, we aim to match either tops for a given bottom or bottoms for a given top. And (2) they did not set the criterion for filtering out users with limited interacted items. Accordingly, we derived our dataset from IQON3000. In particular, we only retained the outfits that contained a top and a bottom, and users who had interacted with no less than 10 and no more than 200 outfits to keep the dataset relatively balanced. Finally, there were 82, 079 user-top-bottom triplets involved 1, 769 users. The detailed statistics are summarized in Table 6.1. The attributes and their corresponding value examples of the derived dataset are shown in Table 6.2.

Notably, all these retained triplets are positive, namely, compatible triplets. We then randomly split these user-top-bottom triplets into four chunks: graph construction set, training set, validation set, and testing set, by the ratio of 6 : 2 : 1 : 1, resulting in a 49, 297 triplet for constructing the heterogeneous graph, 16, 416 triplets for training, 8, 208 triplets for validation, and 8, 208 triplets for testing. Thereafter, as to each positive triplet in the training set, validation set, or testing set, we randomly selected an item (either the top or the bottom) from this triplet as the given item, leaving the other item as the target (positive). Following

Table 6.1 Statistics over our newly constructed dataset based upon IQON3000

Data	Count
User	1, 769
Top	53, 092
Bottom	41, 157
Attribute	98
Outfit (top-bottom)	81, 937
Triplet (user-top-bottom)	82, 079
User historical interacted outfits-min	10
User historical interacted outfits-max	200
User historical interacted outfits-avg	46

that, we replaced the target (positive) item with a randomly sampled item sharing the same category as the target item, to derive a negative triplet. It is worth noting that to ensure fairness, considering the baseline methods do not need the specific graph construction set, we trained them with both the graph construction set and training set, where the negative triplets in the graph construction set were derived in the same manner.

Evaluation Tasks and Metrics

Similar to previous studies [1, 3, 24–26], we justified our proposed MG-POCM scheme via the compatibility estimation task. This task is to evaluate the compatibility score of an arbitrary top-bottom pair for a specific user, where we adopted the AUC (area under the ROC curve) [27] as the evaluation metric. In addition, we evaluated the performance of our model with the complementary item retrieval task [1]. For each positive triplet, we derived a corresponding negative triplet by randomly replacing one item (either a top or a bottom). We merged the original replaced item in the positive triplet and the newly added item in the negative triplet as the set of item candidates. These item candidates were ranked according to their compatibility scores to the given user and item, i.e., the unchanged item in the original positive triplet, calculated by Eq. (6.8). To measure the complementary item retrieval task, we utilized the mean reciprocal ranking (MRR) [28] as the metric.

Implementation Details

Pertaining to the visual embedding of items, we utilized ResNet18 and converted each item image into a 512-D vector. Notably, ResNet18 was also fine-tuned with the whole model. Regarding the textual feature extraction of items, we implemented BERT[3] for Japanese text considering our dataset is in Japanese and embedded each item's textual description into a 768-D vector. The dimension of the final item embedding was $D = 512$. Similarly, using this BERT implementation, each semantic attribute was also embedded into a 768-D vector.

Table 6.2 Attributes and their possible value examples in the derived dataset

Attribute	Possible value examples	Total number
Color	Gray, Black, Red, $\cdots$	12
Price	Low, Middle, High.	3
Category	Coat, Skirt, Jacket, $\cdots$	12
Variety	Tops, Dress, Trousers, $\cdots$	5
Material	Fur, Leather, Denim, $\cdots$	31
Pattern	Stripe, Print, Dot, $\cdots$	15
Design	Frill, V-neck, Ribbons, $\cdots$	13
Dress length	Short, Middle, Long.	3
Sleeve length	Sleeveless, Long, Short, $\cdots$	4

---

[3] https://huggingface.co/cl-tohoku/bert-base-japanese-char/tree/main.

We set the number of layers of all MLPs used in our scheme as 2 and employed Gaussian error linear units (GELU) as the activation function. In practice, we adopted the following set of user-oriented metapaths $\mathcal{P}_{user} = $ {UIAIU, UIU, UIA}, and item-oriented metapaths $\mathcal{P}_{item} = $ {IAI, IUI, II, IIA}. During the subgraph construction for each user/item entity, for efficiency, we set the maximum neighbor size of each node as 5. As to the optimization, we adopted the adaptive moment estimation method (Adam [29]). The learning rate was prepared in the 6% steps to the peak value, which were set to 1e-4, and then linearly decayed to 0. The hyperparameter λ was set to 1, and the batch size was set to 24. All the experiments were implemented by PyTorch over a server equipped with 4 A100-PCIE-40GB GPUs.

6.4.2   Model Comparison

To validate the effectiveness of our proposed scheme, we chose the following state-of-the-art baselines for comparison.

- **GP-BPR** [1] is a comprehensive personal preference modeling scheme, where the multimodal data (e.g., the image and text description) of fashion items are jointly explored.
- **PAI-BPR** [3] is an attributewise interpretable compatibility modeling scheme, which solves the problem of interpretability in clothing matching by locating the discordant and harmonious attributes between fashion items.
- **HFGN** [2] refers to a hierarchical fashion graph network, which simultaneously models the relationships among users, items, and outfits.

Table 6.3 shows the performance comparison among different methods on IQON3000 dataset in terms of AUC and MRR. From this table, we make the following observations. (1) our proposed MG-POCM scheme consistently outperforms all baseline methods over different metrics. In particular, MG-POCM performs better than PAI-BPR and GP-BPR, which indicates the advantage of our scheme that organizes the various entities and relations in the context of POCM into a unified heterogeneous graph and utilizes the metapath-guided heterogeneous graph learning towards personalized outfit compatibility modeling. (2) Our method surpasses the heterogeneous graph-based method HFGN remarkably over both metrics, implying the necessity of considering the items' attributes. (3) GP-BPR outperforms HFGN, which may be because HFGN only utilizes the visual information of fashion items, while GP-BPR jointly considers the images and textual descriptions of items.

6.4.3   Ablation Study

To verify the importance of each component in our model, we conducted ablation experiments on the following derivatives.

Table 6.3 Performance comparison between our proposed MG-POCM and other baseline methods in terms of AUC and MRR over IQON3000. The best results are in bold, while the second best results are underlined

Approaches	PAI-BPR	HFGN	GP-BPR	**MG-POCM**
AUC	0.6096	0.6783	<u>0.7146</u>	**0.7730**
MRR	0.5456	0.6173	<u>0.6346</u>	**0.6427**

Table 6.4 Ablation study of our proposed MG-POCM on IQON3000 dataset. The best results are in bold

Method	AUC	MRR
w/o text	0.7630	0.6392
w/o image	0.7655	0.6382
w/o attribute	0.7339	0.5717
w/o (II,UIA)	0.7639	0.6392
w/o contrastive	0.7647	0.6338
w mean pooling	0.7627	0.6392
MG-POCM	**0.7730**	**0.6427**

- **w/o text**: To study the impact of the textual description of fashion items in POCM, we removed the textual embeddings of items, and kept other parts unchanged.
- **w/o image**: Similarly, to justify the necessity of incorporating the item images in the context of POCM, we omitted the items' visual embeddings, and kept other parts unchanged.
- **w/o attribute**: To verify the importance of the item attributes, we discarded the attribute entities as well as the attribute-related metapaths. The rest of our MG-POCM was unchanged.
- **w/o (II,UIA)**: To validate the necessity of incorporating the metapaths II and UIA, which can be treated as the subpaths of metapaths i.e., IIA and UIAIU, respectively, we omitted them during our heterogeneous graph learning.
- **w/o contrastive**: To explore the effect of the contrastive regularization component, which is used to enhance the latent representation of each entity, we removed the contrastive regularization by setting $\lambda = 0$ in Eq. (5.13).
- **w/ mean pooling**: To evaluate the function of the transformer component in the semantic embedding fusion, we replaced the transformer component with the mean pooling function.

Table 6.4 summarizes the ablation study results. From this table, we observed that our model consistently outperforms all the above derivatives, which demonstrates the effectiveness of each component in our proposed MG-POCM. Specifically, we make the following detailed observations. (1) Both w/o text and w/o image perform inferior to MG-POCM, which suggests that it is essential to consider both modalities of fashion items to boost the item representation learning. (2) w/o attribute presents the worst performance, reflecting the benefit of incorporating the attribute entities as well as their semantic contents into personalized outfit compatibility modeling. (3) w/o (II,UIA) performs worse than our MG-POCM, which suggests subpaths may emphasize the high-order correlations without adding noisy information. (4) The performance of w/o contrastive drops by a large margin, as compared to MG-POCM, indicating that the contrastive regularization is indeed helpful to strengthen the fashion entity representation learning. (5) w/ mean pooling also performs worse than our MG-POCM, reflecting the effectiveness of the transformer in fusing the unfixed number of semantic-enhanced embeddings of users/items.

6.4.4   Sensitivity Analysis

In this part, we evaluated the sensitivity of our model in terms of the number of transformer and GAT layers. In particular, we varied the number of transformer layers from 1 to 5 with the step size of 1. Considering that most of our metapaths involve more than 2 entities, we changed the number of GAT layers from 2 to 6 with the step of 1. Figure 6.5 a and b illustrate the performance of our model on the validation set and testing set with different numbers of transformer layers and GAT layers, respectively. As can be seen, our model achieves relatively stable performance with different numbers of transformer and GAT layers, which implies that our model is not sensitive to these two hyperparameters. Accordingly, in practice, to improve the model efficiency, we set the number of transformer and GAT layers as 1 and 2, respectively.

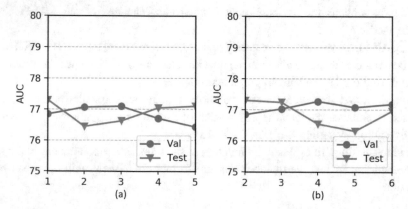

Fig. 6.5 Sensitivity analysis of our model performance in terms of AUC with respect to **a** the number of transformer layers, and **b** the number of GAT layers

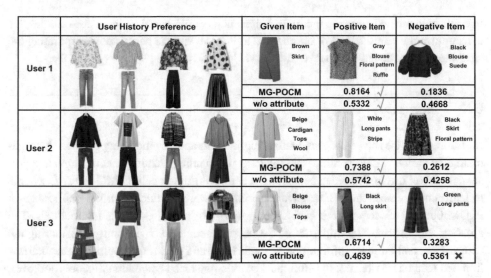

	User History Preference	Given Item	Positive Item	Negative Item
User 1		Brown Skirt	Gray Blouse Floral pattern Ruffle	Black Blouse Suede
		MG-POCM	0.8164 ✓	0.1836
		w/o attribute	0.5332 ✓	0.4668
User 2		Beige Cardigan Tops Wool	White Long pants Stripe	Black Skirt Floral pattern
		MG-POCM	0.7388 ✓	0.2612
		w/o attribute	0.5742 ✓	0.4258
User 3		Beige Blouse Tops	Black Long skirt	Green Long pants
		MG-POCM	0.6714 ✓	0.3283
		w/o attribute	0.4639	0.5361 ✗

Fig. 6.6 Illustration of several POCM results obtained by our MG-POCM and w/o attribute derivative

6.4.5 Case Study

To gain more intuitive insights into our model, we also conducted a case study of our method and the w/o attribute derivative. Figure 6.6 shows three testing samples, where the users' historical preferred top-bottom pairs and items' attributes are also listed to facilitate the experimental result analysis.[4] As can be seen, for the case of the first user with the given brown skirt, although both our MG-POCM and its derivative w/o attribute gives the correct prediction, our MG-POCM assigns a much higher score to the positive item than the negative item. By contrast, w/o attribute gives the former a slightly higher score than the latter item. Namely, our model has a high confidence than its derivative. This may be because incorporating the attribute entities in FPCM enables our model MG-POCM to learn the "floral pattern" that the user prefers for tops, and accordingly gives the positive item with the "floral pattern" a higher score. Similarly, the same phenomenon can be observed in the second case. As can be seen, the second user tends to prefer bottoms of the category "long pants" to match tops, which can be more easily captured by our MG-POCM rather than the w/o attribute derivative. Additionally, as the negative item is a black skirt, which does not go well with the beige cardigan, our MG-POCM assigns a much higher score to the positive item, while w/o attribute only rates a slightly higher score to it, as compared with the negative item. In the last case, we can see that the third user prefers "long skirts" with blouses. The positive item is a black long skirt, looking like long pants, while the negative item is long pants looking like a long skirt. Then with the help of their category attributes, our MG-POCM correctly selects the compatible item for the given top, while the

---

[4] Due to the limited space, we did not provide the text description of the items.

w/o attribute method gives the incorrect judgment. Overall, based on these case studies, we can confirm the effectiveness of our method in POCM, and the benefit of incorporating the attribute information in the context of POCM.

6.5   Summary

In this chapter, we solved the personalized outfit compatibility modeling problem by organizing the various fashion entities and relations into a unified heterogeneous graph and presented a novel metapath-guided personalized compatibility modeling (MG-POCM) scheme to learn entity embeddings. Extensive experiments were conducted on the public dataset IQON3000, which demonstrates the superiority of our model over existing methods. The ablation study verifies the importance of each key module, such as jointly considering the text, image, and attribute information of items towards POCM, the contrastive regularization and using a transformer to fulfill the semantic-enhanced embedding fusion. Moreover, experimental results show that our model is insensitive to the numbers of transformer and GAT layers, which enables the model to perform well with fewer parameters.

References

1. Xuemeng Song, Xianjing Han, Yunkai Li, Jingyuan Chen, Xin-Shun Xu, and Liqiang Nie. 2019. GP-BPR: Personalized Compatibility Modeling for Clothing Matching. In *Proceedings of the ACM International Conference on Multimedia*, 320–328. ACM.
2. Xingchen Li, Xiang Wang, Xiangnan He, Long Chen, Jun Xiao, and Tat-Seng Chua. 2020. Hierarchical Fashion Graph Network for Personalized Outfit Recommendation. In *Proceedings of the International ACM SIGIR Conference on Research and Development in Information Retrieval*, 159–168. ACM.
3. Dikshant Sagar, Jatin Garg, Prarthana Kansal, Sejal Bhalla, Rajiv Ratn Shah, and Yi Yu. 2020. PAI-BPR: Personalized Outfit Recommendation Scheme with Attribute-wise Interpretability. In *IEEE International Conference on Multimedia Big Data*, 221–230. IEEE.
4. Ashish Vaswani, Noam Shazeer, Niki Parmar, Jakob Uszkoreit, Llion Jones, Aidan N. Gomez, Lukasz Kaiser, and Illia Polosukhin. 2017. Attention is All You Need. In *Proceedings of the Advances in Neural Information Processing Systems*, 5998–6008. NIPS.
5. Jian Tang, Meng Qu, Mingzhe Wang, Ming Zhang, Jun Yan, and Qiaozhu Mei. 2015. LINE: Large-Scale Information Network Embedding. In *Proceedings of the World Wide Web Conference*, 1067–1077. ACM.
6. Aditya Grover and Jure Leskovec. 2016. node2vec: Scalable Feature Learning for Networks. In *Proceedings of the International ACM SIGKDD Conference on Knowledge Discovery and Data Mining*, 855–864. ACM.
7. Yu Rong, Wenbing Huang, Tingyang Xu, and Junzhou Huang. 2020. DropEdge: Towards Deep Graph Convolutional Networks on Node Classification. In *Proceedings of the International Conference on Learning Representations*, 1–17. OpenReview.net.
8. Sami Abu-El-Haija, Amol Kapoor, Bryan Perozzi, and Joonseok Lee. 2019. N-GCN: Multi-scale Graph Convolution for Semi-supervised Node Classification. In *Proceedings of the Conference on Uncertainty in Artificial Intelligence*, 841–851. AUAI Press.

9. Wenqi Fan, Yao Ma, Qing Li, Yuan He, Yihong Eric Zhao, Jiliang Tang, and Dawei Yin. 2019. Graph Neural Networks for Social Recommendation. In *Proceedings of the World Wide Web Conference*, 417–426. ACM.

10. Huan Zhao, Quanming Yao, Jianda Li, Yangqiu Song, and Dik Lun Lee. 2017. Meta-Graph Based Recommendation Fusion over Heterogeneous Information Networks. In *Proceedings of the International ACM SIGKDD Conference on Knowledge Discovery and Data Mining*, 635–644. ACM.

11. Yizhou Sun and Jiawei Han. 2012. Mining Heterogeneous Information Networks: Principles and Methodologies. *Synthesis Lectures on Data Mining and Knowledge Discovery* 3 (2): 1–159.

12. Yuxiao Dong, Nitesh V. Chawla, and Ananthram Swami. 2017. Metapath2vec: Scalable Representation Learning for Heterogeneous Networks. In *Proceedings of the International ACM SIGKDD Conference on Knowledge Discovery and Data Mining*, 135–144. ACM.

13. Tomás Mikolov, Kai Chen, Greg Corrado, and Jeffrey Dean. 2013. Efficient Estimation of Word Representations in Vector Space. In *Proceedings of the International Conference on Learning Representations*, 1–12. OpenReview.net.

14. Chuan Shi, Binbin Hu, Wayne Xin Zhao, and Philip S. Yu. 2019. Heterogeneous Information Network Embedding for Recommendation. *IEEE Transactions on Knowledge and Data Engineering* 31 (2): 357–370.

15. Chuxu Zhang, Dongjin Song, Chao Huang, Ananthram Swami, and Nitesh V. Chawla. 2019a. Heterogeneous Graph Neural Network. In *Proceedings of the International ACM SIGKDD Conference on Knowledge Discovery and Data Mining*, 793–803. ACM.

16. Xinyu Fu, Jiani Zhang, Ziqiao Meng, and Irwin King. 2020. MAGNN: Metapath Aggregated Graph Neural Network for Heterogeneous Graph Embedding. In *Proceedings of the World Wide Web Conference*, 2331–2341. ACM.

17. Xiao Wang, Houye Ji, Chuan Shi, Bai Wang, Yanfang Ye, Peng Cui, and Philip S. Yu. 2019. Heterogeneous Graph Attention Network. In *Proceedings of the World Wide Web Conference*, 2022–2032. ACM.

18. Jintao Zhang and Quan Xu. 2021. Attention-aware Heterogeneous Graph Neural Network. *Big Data Mining and Analytics* 4 (4): 233–241.

19. Yuying Xing, Zhao Li, Pengrui Hui, Jiaming Huang, Xia Chen, Long Zhang, and Guoxian Yu. 2020. Link Inference via Heterogeneous Multi-view Graph Neural Networks. In *Proceedings of the International Conference on Database Systems for Advanced Applications*, 698–706. Springer.

20. Di Jin, Cuiying Huo, Chundong Liang, and Liang Yang. 2021. Heterogeneous Graph Neural Network via Attribute Completion. In *Proceedings of the World Wide Web Conference*, 391–400. ACM.

21. Kaiming He, Xiangyu Zhang, Shaoqing Ren, and Jian Sun. 2016. Deep Residual Learning for Image Recognition. In *Proceedings of the IEEE Conference on Computer Vision and Pattern Recognition*, 770–778. IEEE Computer Society.

22. Jacob Devlin, Ming-Wei Chang, Kenton Lee, and Kristina Toutanova. 2019. BERT: Pre-training of Deep Bidirectional Transformers for Language Understanding. In *Proceedings of the Association for Computational Linguistics*, 4171–4186. ACL.

23. Petar Veličković, Guillem Cucurull, Arantxa Casanova, Adriana Romero, Pietro Liò, and Yoshua Bengio. 2018. Graph Attention Networks. In *Proceedings of the International Conference on Learning Representations*, 45–61. OpenReview.net.

24. Weili Guan, Haokun Wen, Xuemeng Song, Chung-Hsing Yeh, Xiaojun Chang, and Liqiang Nie. 2021. Multimodal Compatibility Modeling via Exploring the Consistent and Complementary Correlations. In *Proceedings of the International ACM Conference on Multimedia*, 2299–2307. ACM.

25. Xintong Han, Zuxuan Wu, Yu-Gang Jiang, and Larry S. Davis. 2017. Learning Fashion Compatibility with Bidirectional LSTMs. In *Proceedings of the ACM International Conference on Multimedia*, 1078–1086. ACM.
26. Zeyu Cui, Zekun Li, Shu Wu, Xiaoyu Zhang, and Liang Wang. 2019. Dressing as a Whole: Outfit Compatibility Learning Based on Node-wise Graph Neural Networks. In *Proceedings of the World Wide Web Conference*, 307–317. ACM.
27. Hanwang Zhang, Zheng-Jun Zha, Yang Yang, Shuicheng Yan, Yue Gao, and Tat-Seng Chua. 2013. Attribute-Augmented Semantic Hierarchy: Towards Bridging Semantic Gap and Intention Gap in Image Retrieval. In *Proceedings of the ACM International Conference on Multimedia*, 33–42. ACM.
28. Lu Jiang, Shoou-I Yu, Deyu Meng, Yi Yang, Teruko Mitamura, and Alexander G. Hauptmann. 2015. Fast and Accurate Content-based Semantic Search in 100M Internet Videos. In *Proceedings of the International ACM Conference on Multimedia*, 49–58. ACM.
29. Diederik P. Kingma, and Jimmy Ba. 2015. Adam: A Method for Stochastic Optimization. In *Proceedings of the International Conference on Learning Representations*, 1–15. OpenReview.net.

Research Frontiers

7

Thus far, in this book, we studied the task of outfit compatibility modeling, where each outfit involves a variable number of items. In particular, we first identified the prominent research challenges we faced to solve this task, including the multiple correlated modalities, complicated hidden factors, nonunified semantic attributes, and users' personal preferences. To address these challenges, we proposed a series of graph learning theories. In particular, we first presented a correlation-oriented graph learning method for outfit compatibility modeling, which explicitly models the consistent and complementary relations between different modalities (i.e., the visual and textual modalities). Considering that this scheme overlooks the category modality and the intermodal compatibility modeling, we next introduced a modality-oriented graph learning method for outfit compatibility modeling. Beyond these two methods that focus on the coarse-grained compatibility modeling, we then devised an unsupervised disentangled graph learning method to uncover the hidden factors affecting the overall compatibility and fulfill the fine-grained compatibility modeling. Moreover, to fully utilize item-attribute labels, we further developed a partially supervised disentangled graph learning method. Finally, to incorporate the user's personal tastes, we proposed a metapath-guided heterogeneous graph learning scheme for personalized outfit compatibility modeling.

The theories on compatibility modeling presented in this book can benefit plenty of stakeholders in the fashion industry, such as sellers and end-users. Online sellers, with the help of automatic outfit compatibility modeling, can learn how to compose more eye-catching outfits with fashion items and correlate compatible complementary items in the same product webpage for promoting sales. Additionally, they can recommend compatible complementary items for the user based on his or her previously purchased items. For example, if a user historically bought a white T-shirt, the seller can recommend several jeans for the user. The seller can also fulfill a personalized recommendation with the personalized

1

W. Guan et al., *Graph Learning for Fashion Compatibility Modeling*, Synthesis Lectures on Information Concepts, Retrieval, and Services, https://doi.org/10.1007/978-3-031-18817-6_7

compatibility modeling scheme we introduced. End-users, the main beneficiaries, can learn how to dress properly with the proposed automatic outfit compatibility modeling technique without consulting stylists at great expense, and no longer dread seeking compatible items to create outfits. For online shopping, the user experience can be also largely improved.

Although the above studies have shed some light on outfit compatibility modeling, we have to admit that this research line is still in a young and up-and-coming stage. Here we list a few promising future research directions with their corresponding challenges.

7.1   Efficient Fashion Compatibility Modeling

Fashion compatibility modeling is an essential technique for automatic compatible item recommendation. However, there are millions of fashion items on real-world e-commerce platforms, such as Taobao and Amazon. The number of possible outfits grows exponentially with the number of items. Therefore, in real-world applications, it is highly desirable to develop efficient fashion compatibility modeling methods.

To the best of our knowledge, limited research studies have been devoted to efficient fashion compatibility modeling with graph learning, possibly due to the following challenges. (1) Efficient graph convolution. Traditional graph convolution methods focus on propagating messages among graph nodes and updating entity representations simultaneously at the end of graph convolution. Such graph convolution schemes suffer from two key limitations. First, they do not explicitly distinguish the types of entities during information propagation. Second, their convolution is not totally efficient, especially for the heterogeneous graph that involves multiple types of fashion entities (e.g., users, outfits, items, and attributes) and relations (e.g., user-outfit interactions, outfit-item relations, and item-attribute associations). This is because once a type of entity representation is updated, it can be used timely for updating other types of entities. Therefore, how to fulfill the efficient graph convolution is a difficult challenge. (2) Learning to hash, which aims to learn a compact binary hash code for the given instance, has attracted considerable research interest [1, 2], due to its prominent advantages in saving the time and storage costs. Converting the continuous hash representation of each entity to the binary hash code inevitably loses certain information. Therefore, how to retain the original information to the greatest extent during the hash code learning constitutes another challenge.

In the future, we plan to devise an efficient outfit compatibility modeling scheme, which aims to effectively convert the continuous hash representation of each fashion entity to the binary hash code. It will greatly alleviate the storage complexity and the retrieval cost.

7.2   Unbiased Fashion Compatibility Modeling

Although existing methods have achieved compelling progress, they usually suffer from fitting the spurious correlations between the item categories and the compatibility label of the outfit. Specifically, previous studies resort to the popular fashion-oriented online communities, where fashion experts share their outfit compositions publicly, to collect the positive/compatible outfits, and create negative/incompatible outfits with some random sampling strategy. Therefore, the collected dataset tends to have some bias regarding the item categories. For example, the item categories of pants and jeans occur very frequently in the positive outfits, while categories of kimono and slippers appear few times. Then, the models trained on such datasets tend to learn the spurious correlations between the item categories and the compatibility label of the outfit, while ignoring the other factors (e.g., color and pattern) that affect the outfit compatibility. This undoubtedly would hurt the generalization capability of the deep model. Since it is intractable to collect an unbiased dataset without spurious correlations, as various correlations between item categories and the compatibility labels commonly exist in the real world, the potential solution to this issue is to design an unbiased fashion compatibility modeling scheme that can eliminate the harmful effect of bias reside in the training dataset.

In the future, we plan to use the causal inference technique and introduce the causal graph to inspect the causal relationships between multiple modalities (i.e., visual, textual and category modality) and the outfit compatibility label, to fulfill the unbiased outfit compatibility modeling.

7.3   Try-On Enhanced Fashion Compatibility Modeling

Existing methods on fashion compatibility only focus on modeling the compatibility relationship among discrete items in an outfit but overlook the human habits in fashion compatibility evaluation. In fact, people usually evaluate the matching degree of a given outfit from not only the collocation angle (i.e., in a discrete manner) but also the try-on angle (i.e., in a unified manner), as illustrated in Fig. 7.1.

Therefore, incorporating the try-on effect scheme for outfit compatibility evaluation will boost the model performance. However, combining both the collocation and try-on angles is not nontrivial and may be due to the following challenges. (1) A simple solution to try-on compatibility modeling is to evaluate the compatibility through an outfit's real try-on appearance image. However, the real try-on appearance is usually unavailable in practice. Therefore, how to devise an effective try-on representation learning module, which can capture the potential try-on effect of the given outfit based on its composing items' multimodal content information is a difficult challenge. (2) The compatibility of the same outfit evaluated from the collocation and try-on angles is usually intrinsically consistent. In other words, the knowledge learned from one angle can be used for guiding the learning of the other angle.

Fig. 7.1 Illustration of the compatibility modeling from both collocation and try-on perspectives

Therefore, utilizing the latent consistency to seamlessly integrate the collocation and try-on compatibility modeling, to boost the model performance, poses another challenge.

In the future, we plan to comprehensively analyze the outfit compatibility from both the discrete collocation and unified try-on angles. Moreover, the two compatibility modeling angles can obtain mutually enhanced by absorbing knowledge from each other.

References

1. Jorge Sánchez and Florent Perronnin. 2011. High-Dimensional Signature Compression for Large-Scale Image Classification. In *The Conference on Computer Vision and Pattern Recognition*, 1665–1672. IEEE.
2. Andrea Vedaldi and Andrew Zisserman. 2012. Sparse Kernel Approximations for Efficient Classification and Detection. In *Conference on Computer Vision and Pattern Recognition*, 2320–2327. IEEE.